Project Benefit Management

Project Benefits Management
Linking your Project to the Business

Dr Trish Melton
Dr Peter Iles-Smith
Jim Yates

www.icheme.org

IChemE
heart of the process

AMSTERDAM • BOSTON • HEIDELBERG • LONDON •
NEW YORK • OXFORD • PARIS • SAN DIEGO •
SAN FRANCISCO • SINGAPORE • SYDNEY • TOKYO
Butterworth-Heinemann is an imprint of Elsevier

Butterworth-Heinemann is an imprint of Elsevier
Linacre House, Jordan Hill, Oxford OX2 8DP, UK
30 Corporate Drive, Suite 400, Burlington, MA 01803, USA

First Edition 2008

Copyright © 2008, Trish Melton. Published by Elsevier Ltd. All rights reserved

The right of Author Name to be identified as the author of this work has been asserted in accordance with the Copyright, Designs and Patents Act 1988

No part of this publication may be reproduced, stored in a retrieval system or transmitted in any form or by any means electronic, mechanical, photocopying, recording or otherwise without the prior written permission of the publisher

Permissions may be sought directly from Elsevier's Science & Technology Rights Department in Oxford, UK: phone (+44) (0) 1865 843830; fax (+44) (0) 1865 853333; email: permissions@elsevier.com. Alternatively you can submit your request online by visiting the Elsevier web site at http://elsevier.com/locate/permissions, and selecting *Obtaining permission to use Elsevier material*

Notice
No responsibility is assumed by the publisher for any injury and/or damage to persons or property as a matter of products liability, negligence or otherwise, or from any use or operation of any methods, products, instructions or ideas contained in the material herein. Because of rapid advances in the medical sciences, in particular, independent verification of diagnoses and drug dosages should be made

British Library Cataloguing in Publication Data
A catalogue record for this book is available from the British Library

Library of Congress Cataloging-in-Publication Data
A catalog record for this book is available from the Library of Congress

ISBN: 978-0-7506-8477-4

For information on all Butterworth-Heinemann publications
visit our web site at books.elsevier.com

Typeset by Charon Tec Ltd (A Macmillan Company), Chennai, India
www.charontec.com
Printed and bound in Great Britain

08 09 10   10 9 8 7 6 5 4 3 2 1

Working together to grow
libraries in developing countries

www.elsevier.com | www.bookaid.org | www.sabre.org

ELSEVIER   BOOK AID International   Sabre Foundation

About the authors

Trish Melton is a project and business change professional who has worked on engineering and non-engineering projects world-wide throughout her career. She works predominantly in the chemicals, pharmaceuticals and healthcare industries.

She is a chartered Chemical Engineer and a Fellow of the Institution of Chemical Engineers (IChemE), where she was the founder Chair of the IChemE Project Management Subject Group. She is a part of the Membership Committee which reviews all applications for corporate membership of the institution and in 2005 she was elected to the Council (Board of Trustees).

She is an active member of the International Society of Pharmaceutical Engineering (ISPE) where she serves on the working group in charge of updating ISPE's *Active Pharmaceutical Ingredients (API) Baseline® Guide*. She is the founder and Chair of the ISPE Project Management Community of Practice formed in 2005. She has presented on various subjects at ISPE conferences including project management, quality risk management, and lean manufacturing and has also supported ISPE as the conference leader for project management and pharmaceutical engineering conferences. She is also the developer and lead trainer for ISPE's project management training course. In 2006 the UK Affiliate recognized Trish's achievements when she was awarded their Special Member Recognition Award.

Trish is the Managing Director of MIME Solutions Ltd., an engineering and management consultancy providing project management, business change management, regulatory, and GMP consulting for pharmaceutical, chemical and healthcare clients.

Within her business, Trish is focused on the effective solution of business challenges and these inevitably revolve around some form of project: whether a capital project, an organizational change programme or an interim business solution. Trish uses project management on a daily basis to support the identification of issues for clients and implementation of appropriate, sustainable solutions.

Good project management equals good business management and Trish continues to research and adapt best practice project management in a bid to develop, innovate and offer a more agile approach.

Peter Iles-Smith has 20 years in manufacturing automation projects in the oil and gas, chemical and pharmaceutical industries from both the vendor and user perspective. The last 5 years have been spent developing strategies projects and technologies to improve pharmaceutical manufacturing.

Peter is a chartered Chemical Engineer and founder member of the IChemE Project Management Subject Group and a former Chair of the Process Management and Control Subject Group. He is a founder committee member of the ISPE Community of Practice on Project Management and Education Chair of the World Batch Forum.

About the authors

Jim Yates has worked in project development, predominantly in the process contracting sector, for 30 years, with specific experience in fine chemicals, food and pharmaceuticals. He is a fellow of the Institution of Chemical Engineers and a former Chairman of the North Western Branch.

He currently runs his own management consultancy, Fulcrum Management Limited, providing individual, team and organizational development services across the public and private sectors. He also tutors on the Open University Business School MBA Programme.

Jim's major interests are in linking strategy formation to implementation, enhancing organizational learning and facilitating team development.

About the Project Management Essentials series

The Project Management Essentials series comprises four titles written by experts in their field and developed as practical guidelines, suitable as both university textbooks and refreshers/additional learning for practicing project managers:

- Project Management Toolkit: The Basics for Project Success.
- Project Benefits Management: Linking Projects to the Business.
- Real Project Planning: Developing a Project Delivery Strategy.
- Managing Project Delivery: Maintaining Control and Achieving Success.

The books in the series are supported by an accompanying website http://books.elsevier.com/companions, which delivers blank tool templates for the reader to download for personal use.

Foreword

This book has become a reality for a number of reasons:

- As an experienced Project Manager I realized that more and more I was dealing with customers, sponsors and project team members who had no project management experience. My first book in this series: *Project Management Toolkit*, was a direct response to that. However, I have found that benefits management is a particular area where expertise is lacking, as evidenced by the number of projects which are started with no link to the organization into which they must eventually 'fit'.
- As the founder Chair of the IChemE Project Management Subject Group (PMSG) and then more recently a part of the Continuous Professional Development (CPD) and Publications Sub-groups it was also evident that there wasn't a full series of books which would support the further development of the Project Manager.

Project Benefits Management: Linking your Project to the Business, is intended to be a more in-depth look at the first and final value-added stage in a project and builds on from Chapters 3 and 6 of the *Project Management Toolkit* (Melton, 2007).

The other books in the project management series, developed by the IChemE PMSG, are outlined earlier (page vii).

Although this book is primarily written from the perspective of engineering projects within the process industries, the authors' experiences both outside of this industry and within different types of projects has been used extensively.

The tools, methodologies and examples are specific enough to support engineers managing project benefits within the process industries, yet generic enough to support the R&D manager in understanding the benefits case for the development or launch of a new product; the Business Manager in rationalizing why he should transform a business area, or the IT Manager in proving realization of the benefits of a new computer system.

Projects are achieved for, and by, people. Projects create assets (both 'soft' and 'hard') which are intended to be used by people. Project management must therefore take account of those people whose lives are affected if the assets are not used effectively otherwise, the project will have failed even if it has delivered all its objectives.

This book demonstrates the need for considering, involving and empowering people at a crucial stage in a project – the start! It ensures that 'no project is an island'.

Dr Trish Melton

Acknowledgements

In writing a book which attempts to go into greater detail and to share a greater level of expertise than previously (*Project Management Toolkit*), you need to effectively develop that expertise – gain peer review of that expertise and then share and test it. I therefore want to acknowledge a number of people against these specifics:

For supporting the growth of our projects benefits management expertise over many years:

- All past colleagues and clients.

For supporting the peer review of this collated benefits management expertise:

- Mike Adams (member of the IChemE Project Management Subject Group Committee).
- Bill Wilson, Astrazeneca.

For sharing and testing this collated benefits management expertise on real 'live' projects:

- All current clients, in particular Paul Burke, Bill Wilson and Jeff Wardle, Astrazeneca.
- Associates of MIME Solutions Ltd such as Victoria Bate, Andrew Roberts and others. In particular we want to thank Victoria for her insight and contribution of some unique project challenges.

Finally we would all like to acknowledge the support of our families and in particular to our partners: Andrew, Heather and Elaine – the reason why we start these 'projects' in the first place.

Authors Note: Although all the case studies presented in this book are based on real experiences they have been suitably altered so as to maintain complete confidentiality.

How to use this book

When you pick up this book I am hoping that before you delve into the content you'll start by glancing here.

The structure for the book is based around the benefits management lifecycle which is described in Chapter 1.

Chapter 1 is a general introduction to the concept of benefits management. This can be read at any time to refresh you on some basic concepts which are applied within the core chapters. This chapter also provides the link between the *Project Management Toolkit* (Melton, 2007) and this more in-depth look at the first and fourth value-added stages in a project.

Chapter 2 addresses the more specific organizational issues regarding the need for a robust benefits management process and builds on the basic concepts in Chapter 1.

Chapters 3–6 are the 'core chapters' made up of the following generic sections:

- Introduction of detailed benefits management concepts.
- Presention of specific methodologies and how they support business case development.
- Introduction of benefits management tools and associated tool templates.
- Demonstration of chapter concepts, methodologies and tools, through the use of case studies.
- Summary of handy hints.

Each core chapter can 'stand-alone' and so the reader can dip into any stage of the benefits management process. We have aimed to maintain a logical flow of the concepts but also recognized that in some areas additional amplification, illumination and development was required.

Chapters 7–11 contain a series of case study projects, and in effect are the culmination of the use of all the areas of expertise introduced in the previous chapters. These aim to show the breadth of project benefits management issues that may arise and how these have been dealt with. In addition, a complete example business case is provided for the final case study.

The blank benefits management tool templates are contained on the website http://books.elsevier.com/companions within a protected area. Readers will receive a password with each copy of the book allowing them to access the template. The actual format of the template cannot be changed but the tool can be used electronically by the reader to fill in the project data as required.

And remember...

There will always be someone on your project who is in a great rush to 'start the project' (meaning delivery!) whilst you are pulling together the business case. The hardest job of the Project Manager is harnessing this energy in the right direction.

- Time spent **developing your business case** is more than compensated by a sound understanding of 'why' you need a project, which will support more robust project planning, delivery and ultimately benefits delivery when the project is complete.
- **Start your delivery in haste and repent in leisure** with the mountain of issues which prolong the project life and reduce the chances of success.

    A project needs to go through the full benefits management lifecycle before it's truly complete:
- Without proof of benefits realization it is not possible to state with any confidence that a project was a success.
- A project needs to be integrated back into the business – the goal of a project is to stop being a project and to start 'working' for the business.

Contents

About the authors — v

About the Project Management Essentials series — vii

Foreword — ix

Acknowledgements — xi

How to use this book — xiii

1 Introduction — 1
The project lifecycle — 1
Aims — 2
What is project benefits management? — 3
Why benefits management is needed? — 5
Benefits management path of success — 11
The benefits management lifecycle — 16
And remember . . . — 18

2 Projects and business — 19
Organizational strategy — 19
Developing the organizational strategy and identifying projects — 22
Economic evaluation of projects — 26
Opportunity cost — 26
Link to the business — 27
And remember . . . — 28

3 Benefits concept — 29
Benefits concept development — 31
Tool: Benefits Mapping Tool — 36
Tool: Benefits Matrix — 41
Benefits measurement — 43
Tool: Benefits Scoring Tool — 45

Stakeholder management	48
Benefits realization	50
Handy hints	51
And finally . . .	51

4 Benefits specification – Part 1: linking scope to benefits — 53

Defining project scope	53
Tool: CTQ Scope Definition Tool	55
Challenging the project scope	56
Tool: Scope Challenge Checklist	59
Scope and project strategy	62
Scope and business change management	63
Tool: Business Environment Checklist	66
Handy hints	67
And finally . . .	67

5 Benefits specification – Part 2: business case development — 69

Value-add and lean thinking	70
Cost estimating	72
Benefits specification and measurement	75
Tool: Benefits Influence Matrix	78
How to write a business case	80
Handy hints	85
And finally . . .	85

6 Benefits realization — 87

Realization planning	90
Tool: Benefits Realization Risk Tool	91
Delivery of the explicit benefits	93
Delivery of the implicit benefits	98
Tool: Customer Satisfaction Analysis Tool	100
Benefits sustainability	103
Tool: In Place–In Use Analysis Tool	104
Ending the project – sustaining the business	106
Evaluating success	108
Handy hints	110
And finally . . .	110

7 Short case studies — 111

Case study A – the 'personal' project	111
Lessons learnt	114
Case study B – comparing different approaches to business case development	115

Lessons learnt	117
Case study C – the 'question mark' project	118
Lessons learnt	123

8 Case Study One: product storage and distribution facility project — 125

Situation	125
Potential project	126
Conclusions	130
Lessons learnt	130

9 Case Study Two: pharmaceutical facility refurbishment project — 131

Situation	131
Business case	137
Delivery	146
Benefits realization	147
Conclusions	149
Lessons learnt	149

10 Case Study Three: organizational change programme — 151

Situation	151
Business case	151
Delivery	157
Benefits realization	161
Conclusions	167
Lessons learnt	167

11 Case Study Four: hurricane preparedness project — 169

Situation	169
Business case	170
Delivery	178
Benefits realization	183
Conclusions	186
Lessons learnt	186

12 Appendices — 187

Appendix 12-1 – 'Why?' Checklist	188
Appendix 12-2 – 'Benefits Realized?' Checklist	189
Appendix 12-3 – Economic Evaluation of Projects	190
Appendix 12-4 – 'How?' Checklist	193
Appendix 12-5 – Business Case Template	196
Appendix 12-6 – Simple Benefits Hierarchy Tool	205

Appendix 12-7 – Roadmap Decision Matrix	206
Appendix 12-8 – Benefits Specification Table	206
Appendix 12-9 – Business Case Tool	207
Appendix 12-10 – Benefits Tracking Tool	208
Appendix 12-11 – Sustainability Checklist	209
Appendix 12-12 – Benefits Realization Plan Tool	210
Appendix 12-13 – Glossary	211
Appendix 12-14 – References	212

Index   213

Introduction

This book develops the benefits management concepts originally outlined in *Project Management Toolkit* (Melton, 2007).

The process of planning and delivering a project has typically been the bounded scope of a Project Manager, and therefore of project management. Equally, it has long been recognized that the sole purpose of a project is to work 'off line' from 'business as usual' (BAU) to make some organizational change which ultimately is of benefit to the business. The process of benefits management has been seen as the overall business process which 'fits' the project within the business. Therefore it is clear that benefits management is an integral part of the management of projects and therefore of project management and the project lifecycle.

The project lifecycle

As outlined in *Project Management Toolkit* (Melton, 2007), a project goes through four distinct 'value-added' stages from its start point to its end point (Figure 1-1). Each stage has its own start and end point and each has a specific target to achieve. Effectively each stage can be considered a project within a project.

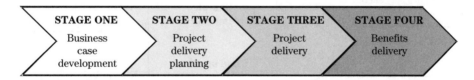

Figure 1-1 The four 'value-added' project stages (Melton, 2007)

Stage One: Business case development

The project start point is usually an idea within the business, for example, an identified need, a change to the status quo or a business requirement for survival. At this stage the project management processes should be challenging whether this is the 'right' project to be progressing.

Stage Two: Project delivery planning

This stage is all about planning and the project management processes are used to determine how to deliver the project 'right.'

Stage Three: Project delivery

Effective delivery is all about the control and management of uncertainty. This stage is therefore focused on the controlled delivery – to deliver the project 'right.'

Stage Four: Benefits delivery

The final stage involves integrating the project into the business – allowing the project to become a part of the normal business process, BAU.

Aims

The aim of this book is to introduce the importance of project benefits management to an audience of Project Managers who have had both good and not so good experiences when developing and delivering their projects. It provides the reader with education; tools and the confidence to manage project benefits so that the chances of successful benefits definition and delivery are increased (Figure 1-2).

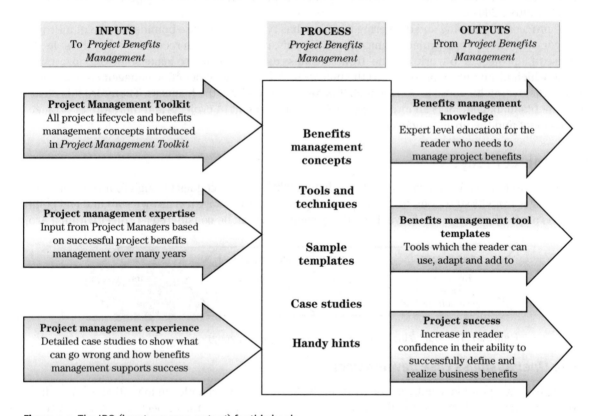

Figure 1-2 The IPO (input–process–output) for this book

Figure 1-2 shows an input-process-output (IPO) diagram for this book and represents the process by which the aims are to be achieved:

- **INPUTS** – lists the inputs to the development of this book.
- **PROCESS** – summarizes the contents of this book.
- **OUTPUTS** – lists the outputs from this book from the perspective of the reader.

Although there are many 'basic' benefits management tools and techniques available, the aim of this book is to introduce methodologies and principles to support the benefits management of more complex projects and programmes. Initially though it will reinforce the project benefits management tools and concepts introduced in *Project Management Toolkit* (Melton, 2007).

- Asking 'why?' is the first value-added stage in a project and the start of benefits definition.
- Asking 'benefits realized?' is the final value-added stage in a project and the way we check benefits delivery.
- Project benefits management links Stage One to Stage Four (Figure 1-1) and ultimately it is the management of the link between a project and the business within which benefits are to be realized.

This book continues to develop generic tools and techniques which can be applied within any type of organization and any type of project. This will be demonstrated in Chapters 7–11 through the use of a variety of different case study projects.

What is project benefits management?

It is clear from many project experiences that there is a poor understanding of benefits management within a project or an organizational context. In some respects benefits management is more aligned to business planning processes than to project management, so it would be expected that managers would be able to look at the strategic management processes within their own organization for guidance. However this is not necessarily the case, benefits management appears to be in its infancy.

The generic benefits management process would typically include:

- *Benefits definition/direction* – working out what benefits types an organization or project is aiming to achieve.
- *Benefits specification* – defining the exact benefits (financial and non-financial benefit metrics) which can be delivered as linked to an organizational activity or project.
- *Benefits realization* – the delivery of the benefits, either during or following completion of the organizational activity or project, as demonstrated through benefit metrics tracking.

Each of the three processes has a direct link to an organizational or project activity, objective, or area of scope. In fact it is this link which differentiates benefits management from other business processes: it aligns both routine and non-routine business activities to the organizational strategy.

- It provides all parts of the business with an understanding of why a specific activity is being done, its importance to the future of the organization and the measurable difference it will make.
- It provides a Project Manager and a Project Team with an understanding of why a project has been approved; its importance to the achievement of the business goals and the benefit metrics which can be tracked to prove that it has delivered the required business change.

Project benefits management can therefore be defined as a business process which links the reason for doing projects with the business impact from their delivery. It sets up the project business case and then proves its successful delivery.

Short case study

A small healthcare organization was growing rapidly and found that there appeared to be many business change projects in progress, with none of them appearing to be progressing significantly. There appeared to be no obvious reason for this. Although many people were involved in many different projects, the overall resource should have been sufficient for both BAU and project work.

Project Benefits Management

In order to understand the situation in more detail a member of the management team was asked to investigate the situation. Following 6 weeks of work the manager reported back to the management team, commenting on the following; data collection, project categories, project flow issues and recommendations.

Data collection

The fact that it took so long to generate an accurate list of all the projects (large and small) that were in progress was a cause for concern. There appeared to be no one route for starting a project. Collecting data on the status of each project was also difficult with varying levels of detail available. However the most shocking part of this phase was the identification of 116 separate projects.

Project categories

The projects were of different types, some involved buying new equipment whilst others were focused on improving current processes. The manager decided to divide the projects into capital funded, operating expense funded or no stated cost, and then sub-divide these into three benefits related categories: strong link to business goals, weak link or no link (Table 1-1). In doing so he was effectively commencing a new benefits management process.

Table 1-1 Example benefits management process

Company resources	Number of projects with a link to business goals		
	Strong	Weak	None
Capital budget	5 (totalling $5,450,000)	4 (totalling $1,150,000)	3 (totalling $150,000)
Operating expense budget	10	25	15
BAU (people, materials)	2	15	37
TOTAL	17	44	55

The total capital budget was easily identified and it was noted that the capital funding approval process appeared to be more robust than revenue expenditure or BAU. However 85% of projects had either no link or at best a weak link to the goals of the business.

Project flow issues

The sheer number of projects in progress at any one time was the main reason that nothing was moving forwards. In addition, some projects had no one clear direction and so were effectively stagnant.

Recommendations

The manager was adamant that the total number of projects needed to be reduced immediately and a project approval system put in place to challenge future projects. However he cautioned against developing too bureaucratic a system which would restrict process improvement at the BAU level.

The management team approved all the recommendations. As a result 75 projects were stopped immediately, 24 were put on hold as low priority at this time, 6 were identified as high priority but needed re-scoping and the remaining 11 were given delivery targets and asked to produce plans to demonstrate what they needed to deliver to these.

The management team also agreed to:

- The development of a project approval system which would ensure alignment of all future projects to their business goals; in effect a benefits management process.
- The development of a clearer set of business goals and use of these in communicating yes/no decisions internally.

Why benefits management is needed?

The above case study describes a basic project benefits management process and in using it, the organization was able to better articulate and manage the benefits they wanted from projects and BAU. In addition it also highlights why benefits management is needed: to focus an organization's limited resources (money, assets and people) on the right things.

To better understand the need for benefits management within a project environment, we have to understand why good project management practices are needed: to prevent chaos at any stage in a project's lifecycle. Project chaos is often described as 'utter confusion' and the symptoms we typically see are:

- Projects delivered late or outside of their agreed budgets.
- Projects that don't deliver to agreed quality, quantity and functionality criterion.
- Projects that don't meet the intended business needs.

It is easy to react to the variety of symptoms but such a reaction can lead to further issues. In order to develop sustainable and robust project management practices the root cause of any symptom needs to be found and resolved (Figure 1-3).

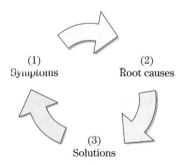

Figure 1-3 Symptoms and root causes

There are many techniques which can be used to identify root causes, but the one used here is 'five whys':

- Ask 'why?' a maximum of five times.
- With each 'why?' the cause becomes more specific and therefore actionable.

- Usually the first/second 'why?' will generate further symptoms.
- Usually the third/fourth 'why?' will generate the cause of the specific project issue.
- Typically the fifth 'why?' will generate the root cause which requires resolution at the organizational level.

Within this chapter (pages 8 and 11) the 5 whys technique has been used to identify project management practices within Stages One and Four of a project that need to be used to deliver project success: the delivery of sustainable business benefits.

Chaos in Stage One – business case development

Examples of typical symptoms of project chaos in Stage One are shown in Figure 1-4. The majority of these are often only seen either during or after project delivery, leading to the conclusion that chaotic or poor business case development impacts the delivery of benefits for an organization.

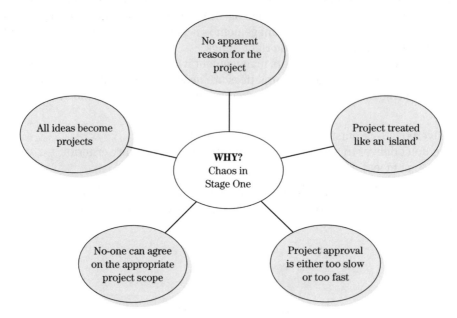

Figure 1-4 Symptoms of project chaos in Stage One 'why?'

It is therefore useful to review key activities at the start of a project which define 'why?' it is required, and what can happen when these are not robustly performed; typically these are:

- Project feasibility work in order to develop a preliminary scope and set of project objectives.
- Benefits specification, so that the defined scope of work and overall project objectives are clearly linked to an agreed, appropriate set of business benefits which can be measured.
- Cost/benefits analysis to challenge whether there is sufficient value in delivering the project.
- Collation of the above and all other business case development activities into a succinct 'deliverable!' – a project business case (which is submitted for organizational approval).

One example of an issue seen at the end of business case development is the amount of time, effort and energy which can be put in to developing a project which then gets rejected. This is demonstrated by the following short case study.

Introduction

Short case study

A project that was expected to go ahead was rejected at the final approval meeting. This resulted in wasted effort, a challenge very late in the project lifecycle and chaos in the Project Team (who were already starting to deliver aspects of the project). The Project Manager was both disappointed and confused and decided to further investigate why this had happened (Figure 1-5) to prevent similar situations occurring again.

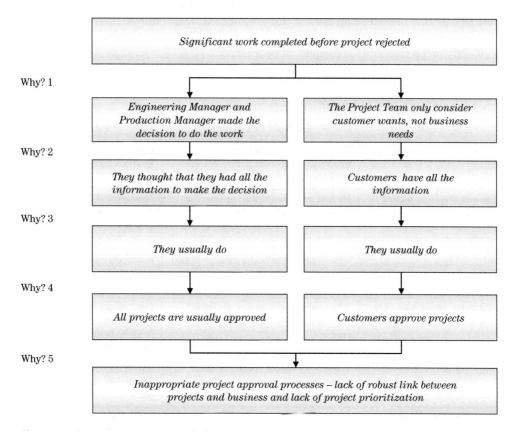

Figure 1-5 Example root cause analysis

In this case, the root cause for the lack of project approval was a disconnect between the project business case development processes and the business for which the project was ultimately intended:

- The business case was developed in isolation of the business climate. It was considered an internal project process.
- The approval processes were slow, allowing Project Teams to develop the business case far beyond what was necessary to really prove that the potential project would deliver benefits if successfully implemented.
- There was little history of project rejection in the organization and therefore poor prioritization processes were used when allocating capital.

Project Benefits Management

As a result of this analysis the organization amended their project approval processes they only allowed a specific expenditure of capital funds to conduct feasibility work as required to develop a robust business case. The project and business management processes were in effect 'reconnected' through a benefits definition process, which supported effective prioritization of capital expenditure across the company.

Generic business case development issues

Table 1-2 shows the results of a few more generic 'five whys' analyses which equally point to an issue in Stage One as the root cause for lack of project success. The common theme through all the 'solutions' is the concept of robust business case development – one theme within this book and the main focus of early project benefits management processes.

Table 1-2 Example root causes – when 'why?' isn't robust

Typical symptoms	Example root causes	Example solutions
No apparent reason for the project	There is no project benefits management process which requires projects to be linked to benefits	Conduct a robust **cost/benefit analysis** against current organizational needs
Project treated like it's an 'island'	Lack of communication between the Project Team and the business	Management of **stakeholders** so that the project remains linked to the business
Project approval is either too slow or too fast	A lack of focus on the project benefits when a project opportunity is identified	Development of a **project benefits management process** to adequately link project approval with benefits required
'Successful' projects not delivering business benefits	A lack of a robust understanding of the benefits which the project needs to enable	Development of a **benefits specification process** to articulate the required benefits and the scope needed to enable them
All ideas become projects	There is no benefit criteria against which to assess if an idea is aligned to business needs	**Specification of benefits** and comparison against organizational goals
No-one can agree on the appropriate project scope	No-one knows why the project is needed	Linking project scope to benefits required by the organization

The solutions above either explicitly or implicitly suggest that the symptoms would be eliminated through the use of a project benefits management process, one outcome of which would be a robust business case. This 'deliverable' is needed to ensure that:

- The 'right' project is delivered – both in its own right and as compared to other projects competing for the limited funds an organization has available.
- A robust specification of benefits is completed allowing a 'baseline' to be measured before the delivery of the project.
- The project has a high potential of success in achieving its outcome.

Chaos in Stage Four – benefits delivery

Examples of typical symptoms of project chaos in Stage Four are shown in Figure 1-6. The majority of these are often only seen either during or after project delivery, leading to the conclusion that chaotic or poor benefits delivery impacts project success.

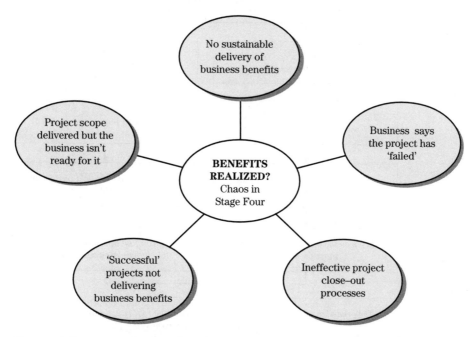

Figure 1-6 Symptoms of project chaos in Stage Four 'benefits realized?'

It is therefore useful to review key activities at the end of a project which support the challenge 'benefits realized?', and what can happen when these are not robustly performed; typically these are:

- Project completion, closure and review to ensure that the project scope has been fully delivered in compliance with the approved business case.
- Benefits tracking to provide evidence that the required benefits metrics are trending towards the target levels.
- Sustainability checking so that any parts of the business which can potentially impact benefits delivery are reviewed and actions put in place to protect achievement of the approved business case.
- Project handover to BAU as a final disengagement for the Project Manager once the project has been integrated, the changes sustained and the benefits realized.

One example of an issue seen during benefits delivery is when an apparently successful project doesn't deliver the anticipated business benefits. This is demonstrated by the following short case study.

Short case study

A project to develop a new electronic system to manage training records had been successfully completed, with the new system fully tested and 'on line' within both schedule and budget targets. However the production department has reported that recent routine audits of the training records had continued to show significant issues and therefore complained that the project had not been successful. The Project Manager was called back in to review and then rectify the situation (Figure 1-7).

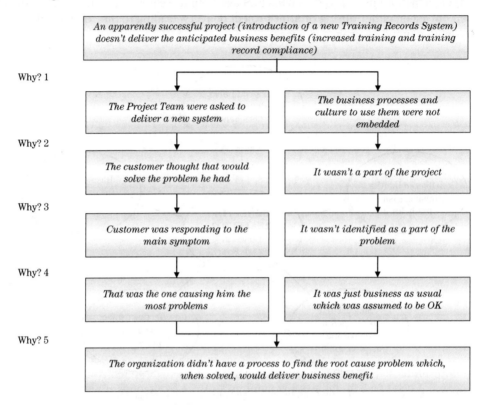

Figure 1-7 Example root cause analysis

In this case, the root cause for the lack of apparent project success was a disconnect between the project and BAU:

- There was no real problem definition which sought to find the root cause of the situation the customer was experiencing, hence there was no understanding of the changes needed to BAU.
- There was no benefits management process to link the project scope to business benefits and therefore the BAU changes required were never identified.
- The Project Team did not consider how the sustainable use of the system (or lack of it) would impact customer perception of success and that it was, in fact, The real business indicator of success.

As a result of this analysis the organization developed a more robust benefits management process which integrated benefits tracking and sustainability checking so that projects were truly integrated into BAU during definition, delivery and close-out.

Generic benefits delivery issues

Table 1-3 shows the results of more root cause analyses which equally point to an issue in Stage Four as the root cause for lack of project success. The common theme through all the 'solutions' is the concept of robust benefits and sustainability tracking – one theme within this book.

Table 1-3 Example root causes – when 'benefits realized?' isn't robust

Typical symptoms	Example root causes	Example solutions
▶ No sustainable delivery of business benefits	The actions to deliver a sustainable business benefit have not been identified	Development of a **benefits realization plan** and a **sustainability plan**
▶ Business says the project has 'failed'	A lack of a robust understanding of the benefits which the project needs to enable	Development of a **benefits realization plan** incorporating **benefits specification** which challenges alignment to the business needs
▶ Ineffective project close–out processes	Integration of the project into BAU is chaotic	Development of a **sustainability plan** to identify the status within and external to the project for handover
▶ 'Successful' projects not delivering business benefits	Success has not been defined to meet the needs of the business	Development of a **benefits realization plan** Understanding customer satisfaction – what will the customer see as success?
▶ Project scope delivered but the business isn't 'ready'	No real understanding of how the project will be integrated within the business	Development of a **sustainability plan** and a **business change plan** to identify the areas outside the project in BAU, which must be in place if the benefits are to be delivered

These solutions either explicitly or implicitly suggest that the symptoms would be eliminated through the development and use of a benefits management process, one outcome of which would be a robust benefits realization plan. This 'deliverable' is needed to ensure that:

- The benefits are specified, targets set and then tracked.
- A sustainability plan is developed to identify and then track areas either within or external to the project scope without which a benefit cannot be sustained.
- A business change plan is developed to identify, deliver and track the required changes to BAU which are potentially outside the scope of the project.
- The project has a high potential of success in meeting its obligations to the business and that those obligations are the right ones.

Benefits management path of success

There is a clear vision of success for the combined activities within Stages One and Four (the elements of the benefits management process (Figure 1-8)) and an associated path of critical success factors (CSFs).

CSF 1 – Business strategy

This CSF is the start point for project benefits management as it sets the business context for all subsequent decisions regarding benefits management. Without a business strategy it is impossible to know whether the projects being considered are valid and value-add for the organization. A clear set of benefits criteria can be generated from a clear strategy and these can be used to find the 'right' project.

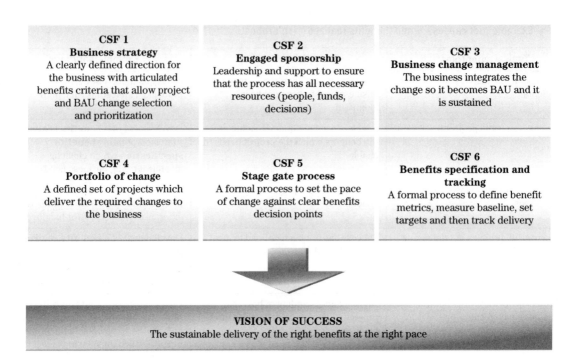

Figure 1-8 Project benefits management success

A key concept of benefits management introduced in *Project Management Toolkit* (Melton, 2007), is the Simple Benefits Hierarchy Tool (Appendix 12-6). A Benefits Hierarchy (Figure 1-9) is a tool which confirms the alignment of the proposed project scope to the targeted business benefits within an organization. It can identify additional benefits as well as highlight gaps or areas of misalignment. At an early stage in the life of a potential project (when the business case is being developed) it is important that the benefits enablers (scope) are defined and accurately linked to the benefits criteria within the organization via the benefits specification.

In effect the benefits hierarchy concept links potential changes within the business to the business strategy and encourages a robust approach when developing a business case.

The hierarchy details five levels of business case development which are critical to project success:

- *Benefits criteria* – this is the reason the project is being done and is an articulation of the organizational goal (Chapters 2 and 3).
- *Business case* – this is the formal specification of the benefits to enable a cost/benefits analysis (Chapter 5).
- *Benefits enablers* – this is the formal development of the scope and the pace of its delivery (Chapter 4).
- *Project objectives and CSFs* – within a business case context this refers to objectives and CSFs definition and project delivery planning (Chapter 4).
- *Benefits realization* – this is the planning, delivery and tracking of specific benefit metrics within an organizational context, linking to business change activities outside of the project (Chapter 6).

In this way a robust business case can be developed so that it is effectively linked to the business. It will contain an appropriate project proposal with scope linked to the delivery of measurable business benefits. This will increase the likelihood of an organization choosing to do the 'right' project.

Introduction

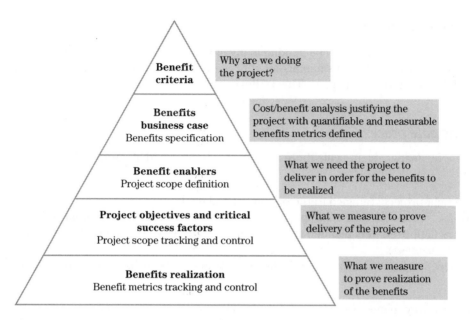

Figure 1-9 The Simple Benefits Hierarchy (Melton, 2007)

CSF 2 – Engaged sponsorship

A successful benefits management process needs effective sponsorship from all levels of the organization (Chapter 6):

- Leaders need to make consistent and appropriate decisions which support the flow of benefits though an organization: selecting the right projects for the right reasons.
- Operational managers and teams need to understand the way that their operations link to the business and generate appropriate data to support benefit metric review.
- Customers (of projects) need to ensure that their interaction with a project is aligned to the benefits case, supporting the release of value to the organization.

CSF 3 – Business change management

Projects are ways to make a step change within a business. A successful project is therefore one which is eventually integrated into the business so the changes become BAU. For example, a project to buy and test a new asset can only release value to the business if:

- The business knows how it can use the asset (increase capacity, new product, new process).
- The business is ready for the asset (facility ready, new business processes developed, training completed).
- The environment is ready (the people working within the business accept the changes and have integrated them into their ways of working).

Business change management and its impact on benefits management is covered in Chapter 4 (scope definition), Chapter 5 (business case) and Chapter 6 (realization of benefits and sustainability of changes).

13

CSF 4 – Portfolio of change

At any one time an organization may be making many changes to the business, requiring a portfolio of projects. The success of the benefits management process is in managing the portfolio so that the 'right' benefits are delivered at the 'right' pace. This is illustrated by a case study (Chapter 10).

CSF 5 – Stage gate process

Within the benefits management cycle there are key questions asked at various stages in the project lifecycle:

- Is this the right project? (aligned to the organizational goal).
- Is this project value-add? (the cost/benefits analysis is appropriate for the organization).
- Is this project appropriately planned to deliver at the right pace? (the benefits will realize at the right time for the business).
- Is this project moving at the right pace? (the benefits are being realized at the right time for the business).

An appropriate way to manage the flow of these decisions is to use a stage gate process. The reality in business is that funds are not limitless. Potential projects (ideas) requiring funding and other organizational resources must be carefully chosen so that they can be successfully delivered and in doing so release the benefits into the business.

Figure 1-10 shows an example of two different stage gate processes for the approval of projects. They demonstrate how two types of project (engineering capital and business change) go through progressive stage gates towards actual project approval. With each successive review additional funds can be released to further progress the idea development rather than a rush to approve the overall delivery of the project. Such a process is critical in keeping the project connected to the business and to the benefits management process within an organization.

For any business this filtering and prioritization of potential projects aligns with the overall concept of project 'front-end loading': spending appropriate time and resources at an early enough stage in a project to:

- Choose the 'right' project to be delivered.
- Increase the potential for success if the project were to be chosen for delivery.

Figure 1-11 summarizes the value of having a stage gate approach. It allows early assessment of ideas which should not be progressed, focusing businesses on the development of value-added activities which deliver the most benefit to the business. It highlights the organizational value of a stage gate approach:

- No organization has limitless resources (time, money and people) and so must only develop projects that will maximize the delivery of benefits to the organization.
- A stage gate approach allows early assessment of ideas which should not be progressed. This focuses the organization on the development of value-added business change activities (on the basis that every project has the potential to change the business in which it is being implemented).
- Every potential project which uses organizational resources can be evaluated to check that it is the 'right' project.

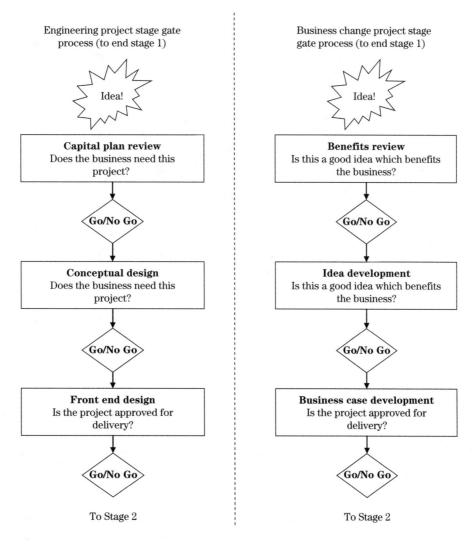

Figure 1-10 An example stage gate approach

A robust business case is a fundamental step in any stage gate process and there is value in allocating resources early in the life of a potential project so that a business case can be developed. This early expenditure of resources supports the appropriate selection of projects for an organization. It is value-add to spend resources on a project which is rejected at an early stage gate on the basis that without the stage gate more resources would have been expended (non-value add).

Note that stage gates would not typically stop at 'project approval' (stage one completion). They are also used to ensure that the chosen project is being appropriately delivered (Chapter 2).

The exact number and type of stage gates within a project approval and management process is typically linked to the project type, funding release processes and benefits management processes. They are usually managed within a portfolio of projects (CSF 4) which can also track any dependencies between projects, activities or benefits. These dependencies can impact stage gate decisions as any project needs to be continually considered as a part of the whole portfolio of change within a business.

Project Benefits Management

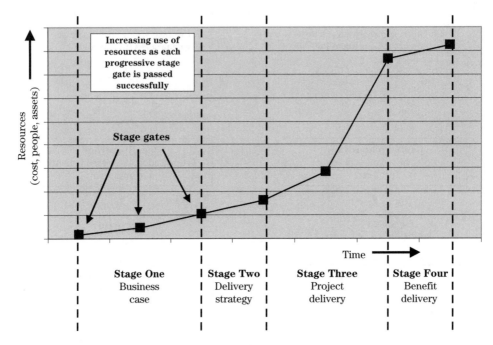

Figure 1-11 The benefits of a stage gate approach

CSF 6 – Benefits specification and tracking

Benefits specification and tracking can occur at all stages in the project lifecycle although typically the following is the case:

- *Stage One* – the business benefits 'baseline' is defined and becomes the basis for the approved business case, where benefits targets are set based on the delivery of a specified scope within a particular business context.
- *Stage Two* – the realization of the benefits is planned in more detail. The benefits specification is set against specific, measurable, achievable, relevant and time-based (SMART) benefits metrics, with a clear understanding of the interaction with other business changes.
- *Stage Three* – the project is delivered and the benefits metrics are tracked throughout. Some change immediately as a result of actually doing the project and others may not be impacted until the full project has been delivered.
- *Stage Four* – the delivery of the business benefits are tracked. Any deviations from plan are investigated and the root causes identified.

The benefits management lifecycle

The benefits management lifecycle (Figure 1-12) is a continuous linkage between the project and the business throughout its life. It is the way that all six CSFs are delivered.

- *Benefits concept* – delivering CSF 1, 2, 3 and 5 in particular.
- *Benefits management* – delivering all CSFs to some extent.
- *Benefits realization* – delivering CSF 3, 4, 5 and 6 in particular.

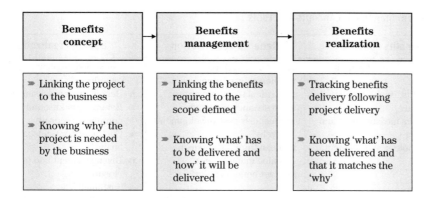

Figure 1-12 The benefits management lifecycle

In this book (Figure 1-13) the two project stages have been divided into the three parts of the benefits management lifecycle.

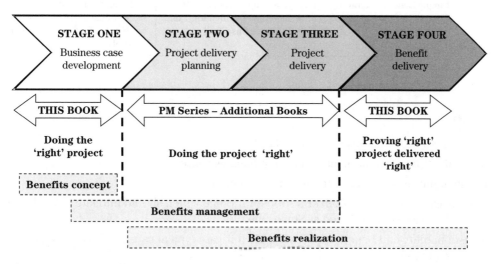

Figure 1-13 Scope of this book

The flow of chapters is put together so that the reader can follow the benefits management lifecycle and cover all CSFs (Figure 1-13). As well as introducing concepts and tools short case studies are used in each chapter to further demonstrate key aspects of benefits management.

Within each chapter additional concepts and tools will be introduced (Table 1-4). In addition benefits management tools from *Project Management Toolkit* (Melton, 2007) and *Real Project Planning* (Melton, 2008) are referenced and included in the appendices for ease of reference.

However, the concept of challenging the benefits management process through using the 'Why?' and 'Benefits Realized?' Checklists remain valid (Melton, 2007) and are a reinforcing concept throughout the case studies (Chapters 7–11).

Project Benefits Management

Table 1-4 Summary of benefits management concepts and tools

	Benefits concept	**Benefits specification**	**Benefits realization**
Concepts	• Development of benefits criteria • Benefits, issues and activity mapping • Benefits measurement and scoring • Stakeholder management	• Scope definition • Hierarchy of objectives • Critical to quality criteria • Resistance to change • Business case development cycle • Lean thinking and lean value management • Cost/benefit analysis	• Delivery of explicit and implicit benefits • Benefits risk assessment • Benefits tracking and cumulative scoring • Customer contracts and Kano analysis • Disengagement and project closure • Sustainability
Tools	• Benefits Mapping Tool • Benefits Matrix • Benefits Scoring Tool	• Scope Definition Tool • Scope Challenge Checklist • Business Environment Checklist • Benefits Influence Matrix • Business Case Template	• Benefits Realization Risk Tool • Customer Satisfaction Analysis Tool • Business Satisfaction Analysis Tool
Reference Tools (Melton, 2007)	• 'Why?' Checklist • Simple Benefits Hierarchy Tool	• 'How?' Checklist • Roadmap Decision Matrix • Benefits Specification Table • Business Case Tool	• 'Benefits Realized?' Checklist • Benefits Tracking Tool • Sustainability Checklist • Benefits Realization Planning Tool

And remember...

- There are four valued-added stages in the project lifecycle.
- Stage One is all about the development of a robust business case.
- Stage Four is all about the delivery of the business benefits.
- Business case development should deliver a robust rationale which should mitigate typical causes of project chaos that can be introduced at Stage One.
- Tracking of sustainability and benefit metrics should mitigate typical causes of project chaos that can be introduced at Stage Four.
- At the end of Stage One it should be clear why a project is needed and how that need relates to a set of specific business benefits. At the end of Stage Four it should be demonstrated that the required benefits have been realized.
- Stage One supports an organization choosing to deliver the 'right' projects and then in setting up those projects for successful delivery. Stage Four proves that the 'right' project has been delivered 'right'.
- The conclusion of Stage One can be checked via use of the 'Why?' Checklist which was introduced in *Project Management Toolkit* (Melton, 2007).
- The conclusion of Stage Four can be checked via use of the 'Benefits Realized?' Checklist which was introduced in *Project Management Toolkit* (Melton, 2007).

2 Projects and business

A project may define a way of reducing cost, removing environmental problems, enhancing the quality of life for a community or determining how to launch a new product, and as such it is not an end in itself, but a means by which the business achieves its objectives.

Benefits management is the process that links those business objectives to the project. Benefits are developed as part of Stage One in the project lifecycle, delivered through Stages Two and Three, then realized in Stage Four.

A key deliverable from the benefits management process is the business case, and to define and develop this effectively, there needs to be an understanding of the underlying business drivers. This requires some understanding of organizational strategy and how it impinges on projects.

Organizational strategy

Organizations have a strategy for four key reasons:

- To achieve a target they have set themselves – their mission.
- To anticipate or respond to changes in the business environment.
- To anticipate or respond to changes in sector competitiveness.
- To satisfy the demands of their key stakeholders.

In most cases, more than one of these generic drivers will contribute to the impetus for the strategy. In many cases several initiatives must come together to form a successful strategy.

Mission

Many organizations have an explicit mission statement and associated vision. In others, the mission of the company is implicit in the organizational culture. Regardless, it will need to have answers to questions such as:

- What business(es) are we in and would we consider going into?
- Where are we trying to get to?
- What are our values, what is important to us?
- How do we want to grow?
- What will be our basis of competition?

A comparison of where it currently is and where it wants to get to will determine the steps needed to achieve its goals. Other aspects will determine how these steps will be taken. A company's mission might be to become the UK's main supplier of a particular chemical. This may require increased capacity which could be satisfied by either a capital development programme or by acquisition. So the mission could be a key driving force for a series of projects.

Project Benefits Management

Changes in the business environment

Today's world is highly dynamic and businesses need to be aware of and able to predict or respond to changes and trends. These global changes affect everyone to some extent but individual changes may have a disproportionate effect on particular industry sectors and even on particular companies.

Effective organizations scan their business environment to identify trends and adjust their business to suit the changes. A number of simple tools are available to help businesses identify trends and scan their environment. Most have developed from the original PEST or STEP model which looked at Political, Economic, Social and Technological developments and their implications for business. More sophisticated models add other areas to consider (e.g. Legal or Ethical). The STEEPLE model provides a useful checklist for organizations to identify changes in the business landscape that could potentially affect them so that they can be better prepared – forewarned is forearmed (Table 2-1).

Table 2-1 STEEPLE model

	Meaning	Typical business driver
S	Social trends – including demographics, such as the aging population in the west	New product development to meet changing market demands
T	Technological trends – inventions and innovation within the industry and more generally. This would include both new process developments and the more general impact of technology such as the internet	Introduction of new and improved processes
E	Economic trends – examples here would be the rise of Asian economies, globalization, the expansion of the European Union	Relocation of production, expansion into new geographic markets
E	Environmental trends – this would cover both the rise of environmental awareness as a political force (see next box) and the impact of environmental changes	Desirable or legislative requirements for environmental improvements or to take account of changes resulting from environmental change such as global warming
P	Political trends – this covers both political issues such as international trade and war, the political agenda (what is energizing people) and issues such as the rise of special interest and pressure groups	Changes to production processes or product types as a result of political or public relations pressures
L	Legal trends – this covers legal and regulatory frameworks. Increases in regulation in some areas and deregulation in others would be relevant, as would changes to specific laws and regulations	Response to changes in legislative or regulatory frameworks
E	Ethical trends – this covers developments in corporate and consumer ethics issues such as the fair trade movement, ethical investments and corporate social responsibility	Exiting business areas that are considered unethical

There are strong overlaps between these headings, but the approach does provide a good framework for analyzing what is going on in the world and deciding what to do about it. Any of these trends could become drivers for projects:

- Changing demographics may lead to changes in demand for particular types of drugs.
- New regulations may lead to changes in product quality specifications.
- Globalization may lead to relocation of production facilities.

Competitive environment

Different industries exhibit different competitive landscapes, and the businesses that operate within these industries need to take account of these as well as the direction of change if they are to become or remain competitive. Professor Michael Porter introduced a simple but highly effective model for understanding what drives competition in an industry (Porter, 1980). In the same way that no project can be considered an island (separate from the business in which it operates) no business can be an island. An organization needs to fit within the industry. Therefore selected projects need to be right for the business and therefore the industry and competitive environment within that industry. Porter suggested that five interlinking forces determine this competitive environment (Figure 2-1):

Competitive Force 1 – Entry of New Competitors	Competitive Force 2 – Threat of Substitutes
▶ How easy is it to get into (and sometimes to get out of) the industry? ▶ What are the capital investment requirements? ▶ What the regulatory hurdles? ▶ What capability is required (global supply base etc.)?	▶ Are alternatives available? ▶ Would customers be willing to use them?

Competitive Force 5 – Rivalry Amongst Existing Competitors
▶ How intensive is this? ▶ How many players? ▶ What is their relative market share? ▶ What are their capabilities?

Competitive Force 3 – The Bargaining Power of Buyers	Competitive Force 4 – The Bargaining Power of Suppliers
▶ How powerful are the buyer industries? ▶ How numerous are they? ▶ What is the size distribution? ▶ Are they organized?	▶ How powerful are the supplier industries? ▶ How numerous are they? ▶ What is the size distribution? ▶ Are they organized?

Figure 2-1 Porter's five forces

Stakeholder expectations

Stakeholders are defined as any person or group affected by, or which has a legitimate claim on, an organization. This means that for most organizations stakeholders will include:

- Shareholders or trustees.
- Customers.
- Employees.
- Suppliers.
- Competitors.
- Neighbours (and by implication the environment).
- Government (local, national and regulators).

Project Benefits Management

Each will have different and sometimes competing expectations and objectives, and organizations need to do what they can to balance these conflicting demands. Changes in stakeholder expectations will often provide an impetus for business change. Examples are:

- Changing regulator focus.
- Changing specifications, supply requirements or service level from customers.
- Neighbourhood concerns about noise or traffic flow.
- Changes to the business to protect employees or improve working conditions.

There are considerable opportunities for overlap between these key drivers and most business change will occur as a response to a combination of factors.

Developing the organizational strategy and identifying projects

Traditionally, the next stage in strategy development is to look at the opportunities and threats presented by the changes in the competitive landscape and consider these in the light of the strengths and weaknesses of the organization. This is a SWOT analysis (Figure 2-2), a method to understand the implications of the external environment whilst developing a good understanding of the internal environment. This is crucial in selecting the 'right' projects.

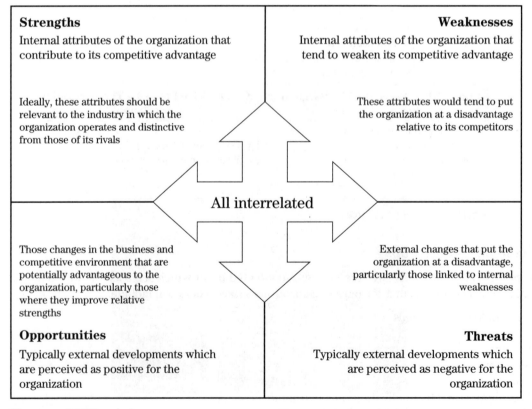

Figure 2-2 SWOT analysis

It is worth noting that whether an external development is seen as a threat or an opportunity is, to some extent, a matter of opinion and attitude.

The organization's strategy is then a balance of its mission, its ability to respond to external developments and its internal capabilities. This would normally result in a range of potential initiatives to deliver the mission, including:

- Mergers and acquisitions.
- Marketing.
- Capital or operating expense funded projects.
- Strategic alliances.
- Change programmes.
- Continuous improvement programmes.

Typically, an organizational initiative will cascade down through different business units and subsidiary companies until specific projects are identified. For example a mission to be the largest manufacturer of a particular product will create an initiative to merge with a competitor, which in turn will create a series of activities to consider; legal (anti-trust and legal entity), financial (valuation, accounting systems), facility (plant asset disposal), marketing (company name, logo, letterhead), and change management (people, positions, business processes) implications. Ultimately there will be a number of projects that need to be completed to support these activities, for example a project to replace the two IT systems with a new common system.

Although project and business objectives need to be aligned, different influences affect them during the life of the project and it is entirely possible for a project to meet its internal objectives but be a business disaster and vice versa. To be fully successful, a project needs to meet both objectives, but it is usually more important that it meets its business objectives than the subsidiary project objectives. Figure 2-3 illustrates the relative importance of achieving project rather than strategic business objectives.

Figure 2-3 Project success

Projects which fully meet their internal objectives but which fail to deliver the required strategic advantages may well be considered 'white elephants'. Those that meet their strategic objectives and deliver an element of competitive advantage to the company will become success stories over time, even if none of the original project objectives were met. Projects which fail to meet either set of objectives are disasters for all concerned.

In most cases, the achievement of a strategic objective does not rely on a single initiative. Even the largest and most complex of projects will require support from other initiatives (such as marketing and culture change programmes) if it is to be successful. Similarly, it is rare for any initiative to contribute to only one objective. Projects are part of the process by which organizations improve their capabilities to compete. There are various means of expressing the links between benefits, projects and corporate initiatives such as the benefits hierarchy, benefits mapping and the benefits matrix (Chapter 3). A project matrix can also be used to show these linkages once the programme of projects to deliver a set of strategic goals has been defined (Chapter 5).

Whilst businesses tend to think predominantly about capital projects, many of the considerations of project selection and project funding are equally relevant to activities such as:

- New product development.
- Skills development.
- Market development.
- Organizational development.

Many of these types of projects are funded from operating expense rather than capital, but the approach to selecting and justifying them should be on the same basis as capital projects.

As we have seen, a project what we are trying to achieve flows from the vision, mission and corporate objectives through a series of local objectives and goals. The project itself provides the means through which these objectives are met. To allow implementation of the project, it is necessary to be very clear about both objectives and the means by which we will achieve them. It is crucial that the objectives are well understood by all participants and that the various elements of the project method or approach are also clear and consistent. It is essential that there are clear and unambiguous statements of:

- The objectives of the project.
- The method to be used to implement it.
- The scope and functionality of the project.
- The budget.
- The programme.

For these to be useful, they all need to be consistent with each other and be prepared on matching bases otherwise it is not possible to be confident in their validity. Where there is ambiguity and conflicting values, the investment decision and supporting business case is likely to be a bewildering process with significant differences between intentions, actions and outcomes.

Ideally, in moving towards the development and approval of the business case, each of these issues needs to be clear and agreed. This allows a relatively easy evaluation of the best course of action. The less certainty in the objectives and approach, the more unstructured decision making becomes. Figure 2-4 illustrates the differences in the approach to organizational decision making under uncertainty. Where there is uncertainty in the objective and the method of achieving it, decision making becomes inspirational (or to be less polite, luck or guesswork). Where the objective is clear (increase production

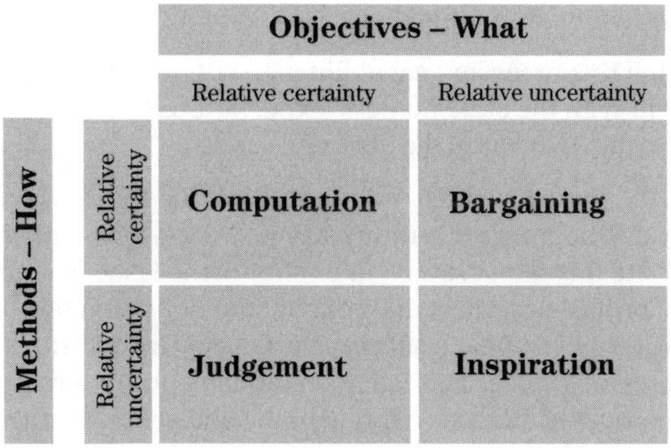

		Objectives – What	
		Relative certainty	Relative uncertainty
Methods – How	Relative certainty	Computation	Bargaining
	Relative uncertainty	Judgement	Inspiration

Figure 2-4 Decision making under uncertainty (Earl and Hopwood, 1980)

capacity by 10%) but the method is not (new plant, de-bottleneck existing plant, process changes) decision making becomes a matter of judgement – pick the best looking idea.

In normal circumstances, the decision route would be to start in the lower right hand box with capital budgets being assigned to a department or company, who would then need to decide how best to spend this money. This needs a degree of creativity or inspiration to identify appropriate and profitable projects. This situation rarely happens because most organizations have at least a partial view of their strategic objectives and how they intend to achieve them.

In most cases, the overall objectives are reasonably clear and once opportunities have been identified, judgment can be exercised to clarify exactly which projects should be undertaken to achieve the objectives.

The top left box is where most Project Team members and managers assume they are. Having a clear understanding of what is to be achieved and the means by which it will be done makes decision making straightforward and allows an analytical/quantifiable approach. Engineers and scientists may feel more at ease with these approaches and this may encourage a predisposition to work on this basis.

The upper right quadrant is typical of 'me too' and 'stay in business' type projects where the overall objectives may be less clear or more difficult to quantify but the nature of the project is well defined. This implies a need for a negotiation led decision making process. A business case developed on this basis will tend to have a very clear method (detailed project implementation strategy) but a 'scattergun' approach to justification in an attempt to satisfy a wide audience. Usually such business cases are developed after the decision to carry out the project has been made and are used to gain access to the necessary funds.

The important feature is that participants should recognize which quadrant they are in as the definition and scope are developed. This will help them to identify where to focus their activities, clarifying the objectives (why are we doing this?) or the methods/means (what are we doing?).

It is worth noting that different Project Team members may well have different perceptions of how clear the means and objectives are. This may lead to problems of focus, attention and motivation and imposes a need on the Project Manager to communicate effectively to close any gaps of understanding.

To try and overcome some of the problems of uncertainty, most sophisticated organizations adopt a formal decision making approach in the form of a stage gate approval (as introduced in Chapter 1). This approach requires that some form of application for approval is made at specific decision stage gates, and the project business case is evaluated relative to the budget available.

Project Benefits Management

Projects need to be reviewed by examining their:

- *Feasibility* – are there the resources and capabilities to complete and exploit them?
- *Acceptability* – do they fit with the stakeholders' expectations?
- *Suitability* – does it help move towards the vision?

At the conceptual stage, the projects which are rejected will be those where the idea does not fit with the key business objectives. At the front end study review, those which will be rejected will be those which do not have an adequate business case. At later stages, the issue is more about confirming the continuing feasibility of the proposed project and refining the definition to maximize benefit.

In some cases, the criteria applied at each successive 'Go/No Go' decision stage gate are more onerous than those at the previous stage. This is understandable for two reasons. Firstly, the costs associated with each stage normally increase exponentially and secondly, the options open to the business are reduced at each stage.

It is important that each stage is treated as a project in itself with clear definition of objectives, deliverables, programme and costs. It may be difficult to apply direct monetary values to the benefits derived from these early project stages, as the real advantage is a reduction in the level of risk associated with the overall project. This results from an improvement in the quality of information, the degree of shared understanding and the elimination of unnecessary features.

Economic evaluation of projects

In making a financial decision to invest in a project, the approach used is to evaluate whether the money invested provides a better return than doing something else with it. The classic comparison is whether the return would be better if the money was left in an interest bearing bank account. Often the method of evaluation and the calculations (typically spreadsheet based) are established by company policy and procedures. The Project Manager needs to understand the basis of the calculations in order to develop the business case.

For operating expense funded projects, which are usually relatively short term, the evaluation tends to be based on the payback period – how long will it take for the investment (of time, resource or money) to be recovered in the savings or benefits of the project?

However, for longer term or high value capital funded projects, sufficient returns are needed to compensate for the lost interest that could have been obtained. It is a fundamental tenet of our economic system that the value of money declines the further in the future we expect to obtain it. Most people would prefer to get £100 today than the same sum in a year's time. Similarly, they would expect to get more than £100 back next year if they were to lend us £100 today. They would need to be compensated by at least the interest they would have earned in the interim.

The return required increases as the time period extends in a geometric manner due to the effect of compounding the interest (money invested at 10% doubles roughly every 7 years), so it is normal to evaluate projects using a discounted cash flow approach. It is important to note that this is not the same as profit. The impact on profit of a capital investment is calculated in a completely different way to its cash flow.

These project economic evaluation techniques are covered in further detail in Appendix 12-3.

Opportunity cost

It is worth bearing in mind that with any investment decision it is only possible to spend the capital once. This means that it is important to consider what else could have been done with the money. The project under consideration should return more value to the organization than the next best alternative. This is the concept of opportunity cost – what do we lose the opportunity to do through making this decision?

However, it is also prudent to take account of the opportunities that the decision enables. This is particularly important in decisions to enter new markets, produce new products and so on.

Similarly appropriate credit should be made for what is enabled in the future by the decision. For example, in one case study (Chapter 7) the organization has an enhanced capability to compete as a result of the project outcome, although it would be unfair to expect the initial capital project to bear the entire cost of the overall programme of change.

In short, it is important to ask: 'What else could we do with this capital and would it give us a better return?' before making any final decision.

Link to the business

Deciding whether or not to execute a project and assigning appropriate budgets/schedules are business decisions. They affect the future survival and prosperity of the organization and need to be given serious consideration. When looking at future events, attention needs to be paid to both the business risks associated with the project and the potential effects of the business environment. The very pressures which lead to projects being selected can lead to major changes during their delivery, particularly if they take more than a couple of years. As has been seen, the world can change very rapidly and it is prudent to plan ahead.

There are a range of approaches to dealing with these issues, from the sensitivity analysis and simulation methods referred to earlier, to longer range approaches such as scenario planning. This allows organizations to identify what approaches they will take under different outcome conditions so that they are properly prepared. Imagine you are expanding capacity on your plant but don't know whether your major competitor will do the same. You may not be able to predict their action but you can plan for either situation.

Spending some time brainstorming and assessing potential risks will be beneficial. The methods used at a project level for risk management can also be applied at the business level. Different management approaches are required for different levels of likelihood and potential impacts. Figure 2-5 illustrates one possible approach.

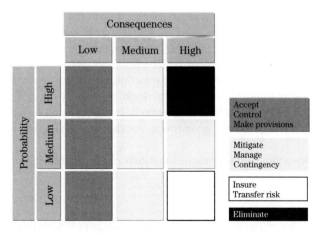

Figure 2-5 Risk management matrix

Projects are business initiatives aimed at meeting strategic goals. The choice of project and its attributes only makes sense in a business context. It cannot be assessed simply from an internal perspective. To make a decision on whether a project is appropriate, there must first be a thorough understanding of the business context.

And remember...

- Projects are one means by which organizations achieve their objectives – they are not ends in themselves.
- Capital or operating expense funding is not limitless and needs to work for the business. Make sure you are spending the right amounts of money on the projects most likely to maximize returns. If the business could earn more by doing a different project (or nothing), then you are doing the wrong project or have not fully understood the benefits.
- Both capital and operating expense funded projects can be complex and involve many people with different views, so gaining a common understanding takes a lot of effort.
- When you are evaluating projects pay as much attention to the benefits as the costs. You may be able to achieve (or claim credit) for more benefits. Increasing the cost may be appropriate if the net effect is to make better returns.
- Make sure that full credit is obtained for features that enable future development, or delete them from the scope.
- Projects often take several years to complete. In today's dynamic business environment the world could be very a different place by the time you have finished.
- Meeting project objectives is not a recipe for success. It helps, but it is more important to meet the business objectives.
- Many projects are rejected because they are under justified rather than because they cost too much. Make sure your project doesn't fail because you haven't put the effort in to identify all of the benefits.

3 Benefits concept

The benefits management lifecycle should be considered a continuous link between the project and the business, throughout the project life (Melton, 2007). It is in Stage One that this link is initially made and it is through identification of benefits criteria that this is possible.

Benefits criteria are strategic business goals which have been translated into a specific benefit category to enable:

- Tracking of specific benefits metrics.
- Definition of the causal relationship between a project objective and a strategic goal.
- Overall performance management of the project at a strategic level.

Benefits criteria describe 'why' a specific project is being delivered and thus form the foundation of the eventual project business case (Chapter 5) and the basis for the project scope (the benefits enablers – Chapter 4). The benefits criteria are at the top of the Benefits Hierarchy (Melton, 2007 and Figure 1-9), which is a tool to demonstrate clear, objective alignment between a potential project, or idea, and a strategic organizational goal.

A project can start from two points:

- From an idea generated from some operational need or issue.
- From a specific desire to achieve a strategic organizational goal.

Typically the former is the case, and it is this that can drive the delivery of projects which have no strategic organizational rationale and no valid business case. To mitigate this, project approval processes are usually in place within organizations to define the link between the project and the business.

Figure 3-1 demonstrates how an operational problem can be systematically reviewed and developed so that it becomes a potential project/idea which:

- Solves the entirety of the operational problem.
- Delivers the maximum strategic benefit for the business.

For example, in a manufacturing facility equipment breakdowns are an operational issue. However, if the facility is already over capacity then the strategic benefit of developing a faster maintenance turnaround time may be minimal.

Figure 3-2 demonstrates how a strategic goal can be systematically reviewed so that a number of potential projects/ideas are developed which:

- Contribute significantly to the achievement of the strategic objective.
- Deliver the maximum strategic benefit for the business.

Project Benefits Management

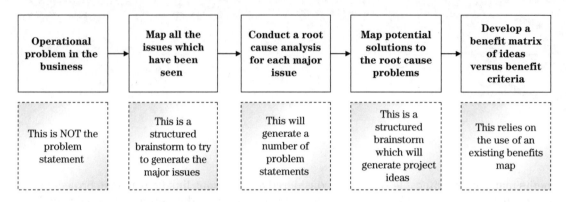

Figure 3-1 From idea to benefits criteria

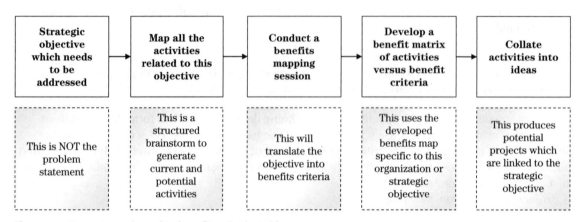

Figure 3-2 From strategic goal to benefits criteria to idea

For example, a manufacturing organization which wants to improve customer satisfaction would be able to understand the relative impact of very different projects such as:

- Improved quality of product.
- Increased delivery speed.
- Reduced service cost so that cost to customer is reduced.
- Improving customer service facilities.

The remainder of this Chapter is concerned with describing the various activities which are required to effectively use Figures 3-1 and 3-2 in order to develop and understand the benefits concept for a project:

- Benefits mapping.
- Benefits measurement.
- Stakeholder management.
- Benefits realization.

As Project Managers have historically not been involved, or necessarily interested, in the approval processes or overall rationale for a project's selection, this chapter illustrates the theme of this whole book – linking the project to the business. Everyone involved with a project should know 'why?' a project is required by the business and also how that project has been selected, possibly above other mutually exclusive projects.

Benefits concept development

The process of generating a benefits map is described in some detail in *Project Management Toolkit* (Melton, 2007). However, a brief summary of the different types and the methodologies for developing a benefits map, and associated maps, are included here.

Benefits mapping

Top-down benefits mapping

In this methodology the benefits map is developed by starting with the organizational objectives. It then develops through a structured brainstorming session by asking 'how can we achieve this?', for each objective, through successive lower level objectives until benefit criteria and benefit metrics are defined.

Figure 3-3 is an example of a benefits map. Here the organizational goal was to have the ability to respond to an increase in sales demand. In asking 'how can we achieve this?' two different organizational objectives were identified: increase plant production and/or increase quality testing capacity. Then the question is asked again 'how can we achieve this?' and benefit criteria are generated. Finally the question 'what is the benefit of this and how do we measure it?' generates benefit metrics.

The top-down approach delivers a set of benefit criteria which can be used by the business to consistently review a portfolio of projects, whether linked or mutually exclusive. This process of evaluating potential projects can then be completed through the development of a benefits matrix (page 37).

Bottom-up benefits mapping

In this methodology the benefits map is developed by starting with the benefits that can be measured and then develops a series of benefits criteria through a structured brainstorming session by asking 'what do we need to achieve this?' The benefits criteria are then challenged by asking 'what will this achieve?' until a robust link to an organizational objective is made.

Figure 3-3 could equally have been developed by a bottom-up approach. Here one benefit metric was defined as an increase in production capacity. The brainstorm questions 'what will we need to achieve this?' In this case it is additional equipment. The next challenge 'what will this achieve?' delivers the organizational or business objective which then delivers the overall strategic goal.

The bottom-up approach often demonstrates that an organization is measuring the wrong things. For instance in Figure 3-3 the traditional company measures of total headcount or equipment utilization do not appear as benefit metrics. Instead, how the people and plant are used to meet the business objective drives the appropriate measures of overall production volume and production volume/person.

The benefits map

In either of the methodologies described:

- Always test the completed benefits map by working back, up or down, through the map to check that the links are robust.
- If there is no robust link between layers in the map then the lower layer will not contribute to the achievement of the strategic goals.
- Ensure that the strategic goals are SMART in themselves (Specific, Measurable, Achievable, Relevant and Time-based).

Project Benefits Management

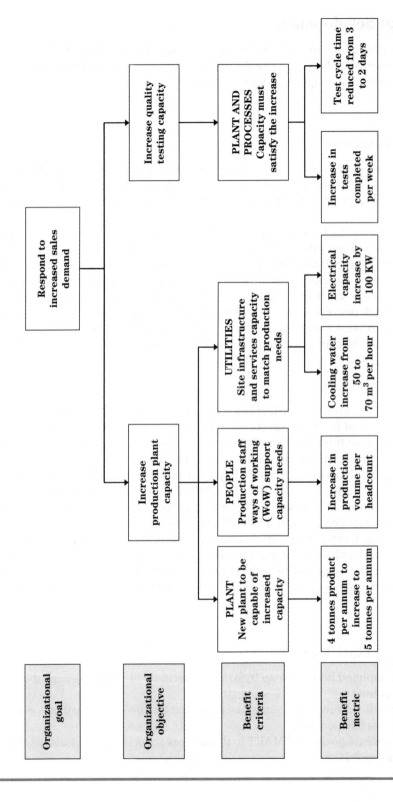

Figure 3-3 Example benefits map

32

Some potential projects may arise out of a desire to achieve a specific business objective and the benefits mapping technique can also be used. If you go 'top-down', ensure there are identified benefits criteria and metrics. If you go 'bottom-up', check there is a robust link to a strategic goal. This methodology forms the basis for the Benefits Mapping Tool (page 46).

Issues mapping

This is a technique which supports benefits concept development and in particular benefits mapping. It is another structured brainstorming methodology which can prevent premature problem definition. There may be many reasons why a situation occurs, and allowing the team review time will help to identify the major issues which have contributed to the current situation. These issues, in themselves, are only symptoms and a formal root cause analysis should be conducted to generate the actual problem statement (the root cause).

Techniques such as 'Fishbone diagrams' are useful to structure brainstorming, as they link cause and effect for the issue under investigation. Figure 3-4 shows a Fishbone or Ishikawa diagram for an operational issue on a manufacturing site. The brainstorm illustrates the likely breadth of causes for product delivery issues.

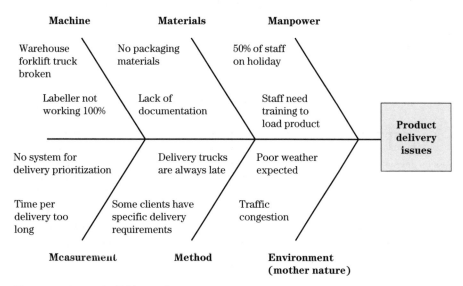

Figure 3-4 Example Fishbone diagram

Alternatively, for a very simple situation, more traditional mind mapping techniques can be used. Figure 3-5 is a very simple example of an issues map generated when a specific supplier on a manufacturing site started to negotiate an increase in cost. The operations team was able to highlight that in this case cost was the least of their problems. There were many other symptoms of supplier performance which didn't match the site benchmark as measured by the site KPIs (key performance indicators).

The root cause analysis for increasing cost (Figure 3-6) then indicated that, in part, the additional cost was to pay for all the things that suppliers should give the site in the first place (good sales service, deliveries on time, efficient handling of queries and fit for purpose raw materials). In addition the cost

Project Benefits Management

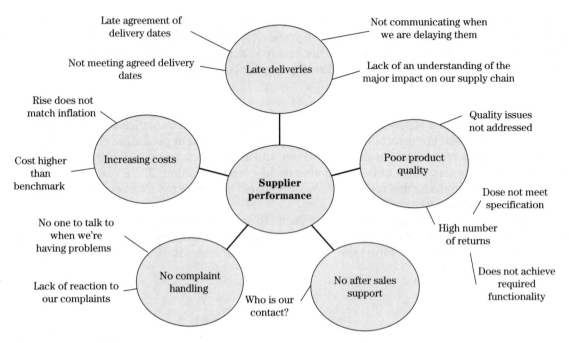

Figure 3-5 Example issues map

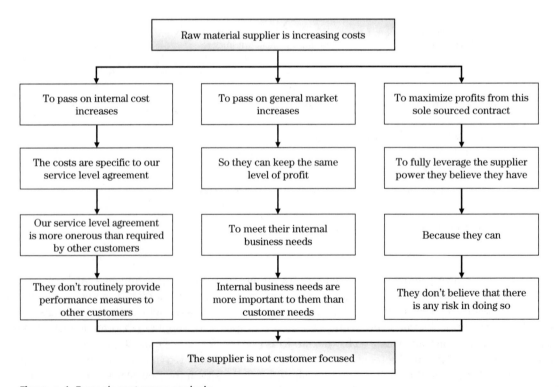

Figure 3-6 Example root cause analysis

Benefits concept

increases were also linked to the suppliers' position as a single sourced provider of a specific raw material. This demonstrated that this was not the type of supplier that was appropriate for the business as it had poor customer focus and was seen to be 'taking advantage' of its preferred supplier status.

So the project became a new supplier sourcing activity rather than a procurement negotiation. The actual project eventually delivered reaped major benefits for the product supply chain (reduced re-work and decreased cycle time) mainly due to the elimination of delivery and quality issues with the specific raw material it relied on.

Activity mapping

This is another structured brainstorming technique which can prevent 'project overload' when strategic objectives are given a high priority. Typically there will already be activities underway (either as distinct projects or part of someone's 'day job') and these need to be identified. However, the 'gaps' also need to be found, those activities which would contribute significantly to the achievement of the strategic goals if they were in progress. Conversely, some activities will have been in progress for some time without any objective articulation of their relationship to a business benefit. This is why an activities mapping session should usually be followed with either a benefits mapping session (bottom-up) or the development of a benefits matrix using criteria generated from a top-down benefits mapping session.

Activities mapping is also useful in identifying dependencies between activities which may have been treated as mutually exclusive. Often apparently lower priority projects can be enablers for larger higher priority projects.

The example in Figure 3-7 is taken from a portfolio of ongoing improvement activities in one area of a research and development business. It highlighted key dependencies between activities and formed the basis of a prioritization process within the business unit. In this example the management team were able to reprioritize the organizational redesign so that the other activities were not impacted. In additional a full review of the benefits from each activity was conducted and the formulation lab upgrade was placed on hold.

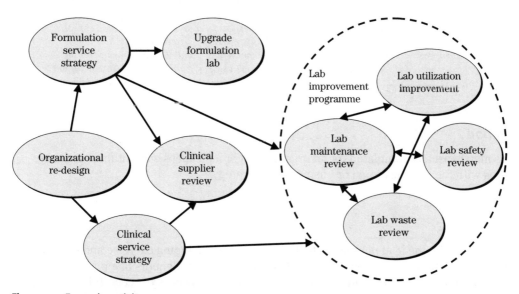

Figure 3-7 Example activity map

Benefits matrix

When an organization has developed a benefits map and an activity map, the next step is to usually connect the two. A benefits matrix is a generic activity which tabulates the identified activities that would deliver various benefits. This type of analysis is often the first step in prioritizing or selecting the ideas/activities which are to be potential projects for development and evaluation. Often the matrix will support the amalgamation of several ideas into one bounded potential project and also allows identification of dependencies. This methodology forms the basis for the Benefits Matrix Tool (page 41).

Tool: Benefits Mapping Tool

The aim of the Benefits Mapping Tool (Table 3-1) is to support teams unfamiliar with the benefits mapping technique (page 31). The aim is to generate a project or organizational benefits map.

Table 3-1 Benefits Mapping Tool explained

Benefits Management Toolkit – Benefits Mapping Tool								
Organization:	<insert area of organization>				**Business Manager:**		<insert name>	
Date:	<insert date>				**Business Sponsor:**		<insert name>	
Top-down development				**Bottom-up challenge**			
Goal	**Objective**	**Criteria**	**Metric**	**Metric challenge**	**Criteria challenge**	**Objective challenge**	**Goal challenge**
<Insert organizational goal>	<Insert how the goal can be achieved>	<Insert how the objective can be achieved>	<insert how the criteria can be measured>	<answer Q1. If we measure this metric what will it tell us?>	<answer Q2. If we achieve this benefit criteria what will we have achieved?>	<answer Q3. If we achieve this objective how will it support the organization?>	<answer Q4. Is this a strategic goal?>

Organization

A benefits map is usually completed for an organization or a business unit within it. It always commences with the organizational or business unit strategy.

Business manager

This would usually be the lead person facilitating the benefits mapping exercise and may be a management team member or an external independent facilitator. Benefits mapping can be a complex exercise and at times it is best externally facilitated, however it is rarely successful if the 'right' team isn't brought together.

Business sponsor

This would need to be the person who leads the organization or business unit. During a benefits mapping workshop this person would usually kick-off the session with an overview of the strategy and ensure that his management team are all in attendance. The Benefits Mapping Tool should enable the management team to adopt clear prioritization processes within their own areas of influence which fully align to the strategic intent. Their engagement in this process is therefore crucial.

Top-down development

This part of the tool follows a typical top-down benefits mapping approach (page 31) and would be done for each strategic goal within the strategy. Occasionally a session is conducted to develop a benefits map for just one, high priority organizational goal but this is only value-add if all activities are to be aligned to this. This is often done for a project or programme aiming to deliver a significant part of a strategic goal.

Bottom-up challenge

Once the benefits map has been developed it is usual to challenge the metrics, criteria and objectives from the bottom-up (page 31). This ensures complete alignment so that the benefit metrics are truly an indication of strategic progress.

Short case study

During a recent site improvement project a benefits mapping session was conducted in order to define the scope of the project: which areas of improvement would deliver the maximum benefit. Cleaning was identified as a concern due to the rising external contract costs coupled with a general decrease in site cleanliness. The benefits mapping session (Figure 3-8 and Table 3-2) highlighted that the main focus of the project should be a change to the site culture as that it didn't get as 'dirty' in the first place. This was quite different to the original idea for cost reduction which was going to be based around a contract review and potential reduction in cleaning standards.

Project Benefits Management

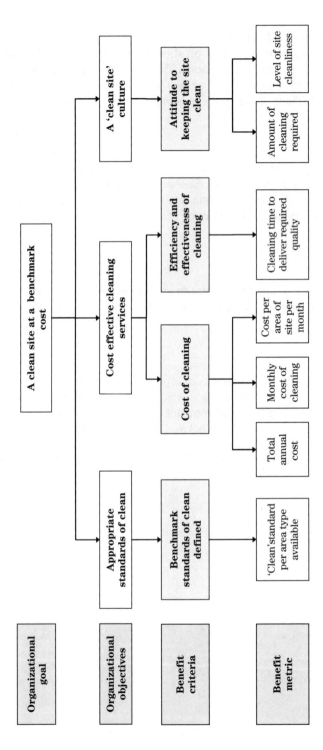

Figure 3-8 Example benefits map

38

Table 3-2 Example Benefits Mapping Tool

Benefits Management Toolkit – Benefits Mapping Tool

Organization:	Site A	Business Manager:	Jane Jones (FM Manager)
Date:	Date x	Business Sponsor:	Sarah Smith (Site Director)

Top-down development				Bottom-up challenge			
Goal	Objective	Criteria	Metric	Metric challenge	Criteria challenge	Objective challenge	Goal challenge
A clean site at a benchmark cost	Appropriate standards of clean	Defined benchmark standard cf clean	'Clean' standard defined per area of site	By having a standard I have the ability to compare this standard to other sites – I will know if it is appropriate	By having a benchmark I will know whether we are setting the appropriate goals for cleaning	By having a cleaning goal which is appropriate I will be able to have a site which is appropriately clean	It is important to the organization that the site is clean – it is part of its reputation and linked to safety and good manufacturing practice (GMP) culture
	Cost effective cleaning services	Cost of cleaning	Total annual cost	By measuring total annual cost I can check it meets planned costs and also benchmark costs	By comparing costs in this way I can determine if the cleaning expenditure is appropriate. Likely needs to trend on a monthly basis	In knowing whether I have spent too much or too little on cleaning I can work out if it has been value-add	It is important that we spend all money wisely
			Cost per area of site	By measuring this ratio I can look at very different parts of the site and compare with benchmarks			
		Efficiency of cleaning	Cleaning effort per clean standard	If I know how much effort it has taken to clean the site to a specific quality level I can check against benchmarks	By comparing effort in this way I can assess the effectiveness of the external service provider	I can assess whether this is the right service provider to give me a clean site – need to consider efficiency and effectiveness	It is important that we work with the right service providers – looking at both cost and quality

(Continued)

Table 3-2 (Continued)

Benefits Management Toolkit – Benefits Mapping Tool

Organization:	Site A	Business Manager:	Jane Jones (FM Manager)
Date:	Date x	Business Sponsor:	Sarah Smith (Site Director)

Top-down development				Bottom-up challenge			
Goal	Objective	Criteria	Metric	Metric challenge	Criteria challenge	Objective challenge	Goal challenge
	A 'clean site' culture	Increased 'clean as you go' attitude	Site cleanliness	By measuring this I know how much 'dirt' is being generated by the people working on the site	If I know how much 'dirt' is being generated I can evaluate how much people are keeping their work areas clean and their attitudes to cleanliness	If the 'clean culture' has improved then we will have a cleaner site and be able to reduce cleaning costs	We want to be a site that generates less 'waste' and by being cleaner we will do that
			Cleaning time	By measuring this I can trend how much effort was required by the cleaners	If the effort reduces then I know that this is likely to be linked to a reduced amount of dirt on the site in the first place (or could be linked to decreased effectiveness or increased efficiency of cleaner)		

Tool: Benefits Matrix

The aim of the Benefit Matrix Tool (Table 3-3) is to evaluate a set of competing ideas, activities and/or projects. It requires a good understanding of the organizational or project benefits map so that it can evaluate:

- Which set of ideas, activities or projects will deliver the maximum benefits, both individually and in combination with each other.
- Dependencies between ideas, activities or projects.
- Ideas, activities or projects which are not mutually exclusive, they may deliver the same benefits (which are not cumulative) or may even cancel out each others benefits.

Table 3-3 Benefits Matrix explained

Benefits Management Toolkit – Benefits Matrix				
Project: <insert project name>			**Project Manager:** <insert name>	
Date: <insert date>			**Project Sponsor:** <insert name>	
Activity	Benefits criteria			
	<insert benefit criteria 1>	<insert benefit criteria 2>	<insert benefit criteria 3>	<insert benefit criteria n>
<insert idea, activity or project>	<insert level of benefit>	<insert level of benefit>	<insert level of benefit>	<insert level of benefit>
<insert idea, activity or project>	<insert level of benefit>	<insert level of benefit>	<insert level of benefit>	<insert level of benefit>

Activity

Within a specific benefit matrix it is usual to start by evaluating activities and/or projects which are in progress and then to add ideas to review gaps in the benefits profile. The list inserted may well have been generated by using an activities mapping technique to start the review of activity dependency and/or benefit dependency.

Benefits criteria

These are the benefits areas generated by benefits mapping against organizational goals. Through prioritizing the goals the benefits criteria can also be prioritized. In this way the quantity of benefit can be reviewed as well as the type. It is usual for some preliminary scoring to be done. For example an activity might deliver a low level of benefit 1 and a high level of benefit 2. Other scoring systems just indicate the main focus of the activity:

- *Major* – the activity has a major contribution to the achievement of this benefit criteria.
- *Minor* – the activity has a minor contribution to the achievement of this benefit criteria.
- *Enabling* – the activity is an enabler of this benefit and therefore needs to be completed prior to any activity being able to deliver a major or minor contribution.

Once the benefits scoring has been completed a decision can be taken on which projects should be in or out of a current portfolio. In this way an organization can be sure that the portfolio of change is appropriate and that each project has a reason for being selected.

Project Benefits Management

Short case study

A manufacturing site is attempting to respond to an increased sales demand for a specific product. In order to understand the most appropriate activities to progress they conducted a benefits mapping session (Figure 3-3) and then looked at four potential projects through use of a benefits matrix (Table 3-4). Each project, when considered individually, appears to deliver some level of financial benefit and typically this is how they would be selected within this organization.

Table 3-4 Example benefits matrix

Benefits Management Toolkit – Benefits Matrix				
Project: Sales improvement programme			**Project Manager:** confidential	
Date: confidential			**Project Sponsor:** confidential	
Activity	**Benefits criteria**			
	Production plant	Production people	Production utilities	Quality lab plant and processes
1. Plant automation	**Major** Increase production	**Minor** Fewer people needed – likely reduce product cost	**None** Insufficient utilities for increased production	**None** More samples waiting to be tested
2. Quality lab process improvement	None	None	None	**Enabling** Lab would have some spare time/capacity as would get through current testing capacity quicker
3. Extended production working hours	**Minor** Increase production to the limits of the current plant – reduced downtime for maintenance	**None** No increase in people efficiency and product cost may even increase	**None** Additional minor production increase uses only spare utility capacity	**None** More samples waiting to be tested
4. Installation of new production equipment (bottleneck areas only)	**Major** Increase production	**Minor** By eliminating the bottleneck the current staff can be used more effectively	**None** Insufficient utilities for increased production	**None** More samples waiting to be tested

By analysing the projects using a benefits matrix the organization were able to identify that:

- The dependencies between projects two and four (an enabling project and one which has obvious benefits).
- That projects one, three and four are not mutually exclusive (so they are not all necessarily needed). Further investigation should show whether the benefits are cumulative or just different ways of delivering the same benefit.
- That there is still one project missing if the overall business objectives are to be met – ensuring that there are sufficient utilities for the increased production.

Looking at a portfolio in this way is an effective start to articulating the benefits specification for a chosen project (Chapter 5).

Benefits measurement

Benefits are measured at various stages during project evaluation and business case development:

- At a very early stage when ideas are being proposed it is enough to find an alignment between an idea and a business benefit (using the benefit mapping and benefit matrix, for example).
- As business resources are expended the level of articulation required to successively release further resources increases:
  - Qualitative benefits metrics definition (define what will be measured).
  - Quantitative benefits scoring (as a comparator measure, not an absolute measure of what will be delivered).
  - Quantitative benefit metrics with a baseline and target measure, over a defined timeline (benefits specification).

Benefits metrics

Establishing benefits criteria is the first stage in defining benefit metrics, which are typically detailed within Specific Benefits Specification Tables (Chapter 5). Defining benefits metrics also supports the development of the project scope (Chapter 4) through the definition of a causal relationship to a key scope area namely a critical success factor (CSF). This is another concept originally introduced in *Project Management Toolkit* (Melton, 2007):

- Those project objectives (sometimes called benefit enablers) which measure areas of scope that are critical to the delivery of the project are called CSFs.
- CSFs enable the delivery of a specific benefit metric or set of benefit metrics.

Benefits scoring

Once benefits criteria have been established and a benefits matrix generated, it is possible to develop more sophisticated techniques to compare projects. This involves the generation of a benefits score which is a comparative rather than absolute measure. Benefits scoring systems are usually linked to a company's performance management system, so that any portfolio of projects is balanced with the overall needs of the organization.

For example, if the general approach of a balanced scorecard (Kaplan and Norton, 1996) is used then an organization will have performance measures in place to monitor the achievement of key goals related to four different perspectives:

- Financial.
- Customer.
- Learning and growing.
- Internal business processes.

A benefits scoring system would weight benefits criteria in each of the four areas equally, so that projects with little financial benefit would still be considered on the basis of achieving other strategic goals.

Project Benefits Management

Other more complex performance management models can be used to support overall business management and improvement. However, they are all based on the basic premise that an organization needs to balance financial and non-financial goals in order to perform sustainably in its chosen arena.

Performance management – the benefits scorecard

A simple benefits scorecard (Figure 3-9) was introduced in *Project Management Toolkit* (Melton, 2007). Such a scorecard would usually be based on a complex organizational benefits map derived from the full set of strategic goals.

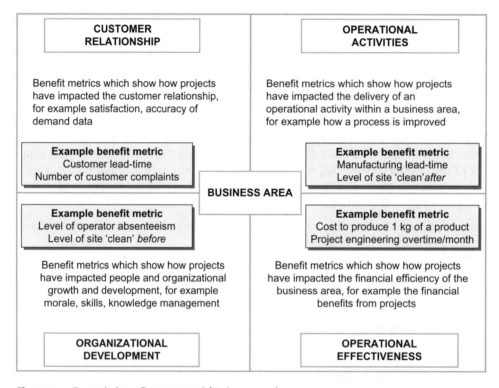

Figure 3-9 Example benefits scorecard (Melton, 2007)

A performance management system, within an organization or project, impacts culture:

- An organization or project takes a measurement of some criteria because it is an input to a decision.
- Decisions cause specific actions.
- Measures cause specific behaviours.
- Culture is a combination of behaviours, actions and values.

It is therefore crucial that the link between measures and culture (expressed as behaviours, actions and values) is determined so that an appropriate culture develops.

Short case study

A Project Manager had developed a performance scorecard to ensure that all schedule milestones were met. At the first project review meeting a non-critical milestone had been missed by a team member and he was publicly admonished for it. In subsequent meetings similar actions were seen and as a result the duration estimates for future work became more conservative and full of 'safety'. The team members developed a culture to protect themselves. However the result was that critical milestones started to be impacted and the project fell way behind schedule even though the team were meeting the deadlines.

Because of the measures in place, and the actions of the Project Manager, the team members behaved in a way that protected themselves. If the measures had been focused on a team goal (such as meeting fewer but critical milestones) where success/failure was shared then the project culture would have been quite different.

Tool: Benefits Scoring Tool

The aim of the Benefits Scoring Tool (Table 3-5) is to support an organization in assessing the cumulative benefits impact of an idea, activity or project. It builds on the work an organization has done in compiling a benefits matrix and so should only be done for ideas, activities or projects that have a high priority alignment to organizational goals. The tool would normally be used as a part of developing the business case after early feasibility work and scope definition has been completed. The method to assign benefits scores uses a standard ranking system whereby benefits criteria are given a weighting relative to each other and then a project can be assessed against its potential to deliver against individual criteria. The scoring system is 'normalized' so that different projects can be assessed and then their scores compared.

Table 3-5 Benefits Scoring Tool explained

Benefits Management Toolkit – Benefits Scoring Tool				
Project:	<insert project name>	**Project Manager:**		<insert name>
Date:	<insert date>	**Project Sponsor:**		<insert name>
Non-financial benefits score				
Benefit criteria	**Maximum score**	**Benefit level**	**Benefit impact**	**Benefit score**
<insert benefit criteria>	<insert maximum score possible based on priority of criteria>	<insert the level of contribution of this project to this benefit criteria>	<insert the organizational impact of the benefit delivery from this project>	<calculate score>
Financial benefits score			**Total benefit score**	
Benefit type	**Benefit level**	**Benefit multiplier**	**Score**	**Comment**
<insert type of financial benefit>	<insert financial value>	<calculate multiplier>	<calculate score>	<insert comment on level of score versus required level>

Non-financial benefits score

In order to calculate a non-financial benefits score the following data is needed:

Benefit criteria

Insert the specific benefit criteria that this project will impact. This will usually have been identified previously within a benefit matrix or benefits map.

Maximum score

Based on the overall benefits profile identified during the benefits mapping session and the current priority for delivery of benefits, there will be a maximum score that can be achieved for each criteria. This effectively gives each criteria a weighting as compared to other criteria. In a three-tier system criteria can be weighted as 5, 10 and 15, for example. Overall the scores are usually normalized so that the total delivery of the benefits criteria equates to a score of 100 for an organization.

Benefit level

As further work will have been completed on the project scope definition, and therefore the benefits specification, the level of benefit should be better articulated then simply major or minor.

- If the maximum score is 15 then it is usual to score the benefits level between 1 and 15.
- The exact level should be equivalent to an increase in some specified metric although this may not have been fully defined before project approval.

The method for the calculation of the normalized non-financial benefit level should be articulated. For example, an increase in customer satisfaction may be measured through a survey and the survey results aligned to a benefits score: an increase by 10% satisfaction being equivalent to a benefits level of 5.

Benefit impact

A project may deliver significantly against a specific criteria but it may only be linked to a specific part of the organization. The scoring needs to take account of this. For example, the cost of cleaning offices and labs may be improved yet this would have no impact on manufacturing facility cleaning costs which represent 60% of the total cleaning expenditure. Therefore the impact score is 40%.

Benefit score

The non-financial benefit score is a calculation based on the previous data input:

- Benefit score = benefit level × benefit impact.

Financial benefits score

In order to calculate a financial benefits score the following data is needed, as well as support from financial managers from within the organization:

Benefit type

There are four main types of financial benefit:

1. **Sustainable financial benefits**
   These are savings which are made year-on-year, for example reduction in overtime %, headcount reduction, increase in operational efficiencies which may also reduce waste and increase the utilization of assets. These are financial savings which impact the current organizational budget that has already been approved, so they are 'real' financial savings from the 'bottom line'.

   For example, being able to deliver an existing product or service with fewer resources should deliver sustainable financial benefits. In order to realize the benefits though, you would have to remove the surplus resources from one area and use them elsewhere. Equally the increase in a product or service quality may also bring sustainable financial benefits which are unrelated to resource management changes.

2. **One-off financial savings**
   These tend to be related to the reduction in use of working capital and are linked to the inventory of materials required to operate. For example less stock of any material is a one-off financial saving.

   Another example would be the reduction in storing maintenance spares. The first time you reduce the stock there are 'real' financial savings, however subsequently you need to sustain the reduced level to merely 'keep' the saving.

3. **Financial cost avoidance**
   These tend to be related to the avoidance of capital expenditure which has already been included in capital budgets for current or future years.

   Some projects can change ways of working (WoW) to an extent that a capital investment is no longer needed. For example the 'lean' use of existing assets through changing WoW (processes, procedures, culture) may mean that a planned equipment purchase is no longer required.

4. **Increase in sales performance**
   Projects may also contribute to the increase in sales performance for any part of the organization and therefore deliver financial benefits to the organization.

Benefit level

The financial value should be indicated here. Often this is the key rationale for a project but in this tool it becomes merely one part of the overall benefits 'picture'.

Benefit multiplier

The organization should identify the level of multiplier per benefit type and level, for example:

- A sustainable financial benefit of less than $50,000 = 1; up to $100,000 = 1.2 and greater than $100,000 = 1.5.
- A one-off financial saving or a financial cost avoidance of less than $100,000 = 1.

The benefit multiplier is a way for the organization to balance financial versus non-financial benefits and may change depending on the current situation within an organization.

Total benefits score

In order to complete this section the following data is needed:

Score

The final score is calculated as follows:

- Score = non-financial benefits score × financial benefits multiplier.

This should give a benefits score which can be compared against other project benefits scores within the same organization.

Comment

At this stage the Project Manager may be able to make a comment on the score relative to the norms within the organization. For example a score of less than a specific number may be low priority. On the other hand sometimes the absolute scores don't tell the full story and the benefits score needs to be assessed in light of the overall portfolio of change. An organization needs to consider:

- 'Balance' in terms of its improvement portfolio linked back to the balanced scorecard approach to organizational performance management. For example, even if all cost reduction projects achieve high scores, this does not mean that they should all be done if it is at the expense of improvement projects in other areas (such as customer relationship management).
- How much change various areas of an organization can handle at any one time.
- Which projects deliver enabling benefits for the organization.

A typical method to compare one project to another in terms of its benefits score is by using a pareto analysis (Figure 3.10), which identified the set of projects that would deliver 80% of the benefits. In the case of the example data in Figure 3-10 the following actions were taken:

- Project A–P would deliver 80% of the required benefits and so these would be resourced as a matter of priority.
- All projects after P, with the exception of those identified as enabling for those in A–P, would be stopped.

Stakeholder management

In any project situation there will be a number of stakeholders who will be either supporting or resisting the potential project. Understanding the links to the strategic goals would usually require a high level of senior stakeholder involvement, so that the benefits criteria and the causal link to high priority strategic objectives can be defined.

Stakeholder mapping techniques should be used to identify who should be involved at this early stage of a potential project:

- *The sponsor* – the person in the organization who will ultimately be accountable for the delivery of the business benefits should the project get approved.

Benefits concept

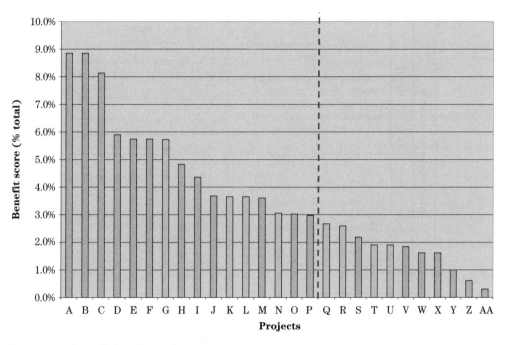

Figure 3-10 Example benefit scoring pareto

- *A project champion* – who may also be the sponsor, but equally may be someone lower in the organization, who supported the identification of the operational issue.
- *The end user* – even at this early stage, the involvement of someone who will be impacted by the project, if implemented, is useful in building support and understanding (however this needs to be at an appropriate level in the organization).

Figure 3-11 shows an ideal situation where all the required support has the appropriate level of importance and influence. Issues can arise if:

- The sponsor is inappropriately chosen and has insufficient authority (a very typical problem).
- A senior stakeholder with power is not behind the project and becomes an active resistor.

Once all stakeholders have been identified and mapped, a plan to manage any issues can be put in place:

- How should sponsors and champions in Boxes 1, 2 and 3 be appropriately involved? How should they be 'kept in the loop'?
- How should the resisting stakeholder in Box 4 be dealt with? Can any sponsors in Box 1 be used to deal with issues? Are there any potential resistors in other boxes which need to be 'moved'?

Stakeholder management supports the identification of the link between projects and the business and stakeholders should be effectively engaged to support the development (and championing) of the resultant business case.

Project Benefits Management

		LOW	HIGH
Stakeholder importance	HIGH	**Box 2 – Champions, end users** Some people may have less influence as they are lower in the organization. However their knowledge and experience needs to be harnessed	**Box 1 – Sponsors, champions** If the sponsor isn't in this box then the potential project may not be able to face the challenges from any resisting stakeholders Generating champions in this box who can support the sponsor is also useful
	LOW	**Box 4 – Resistors** Stakeholders who are likely to resist or strongly challenge the potential project should be in this box	**Box 3 – Champions, end users** The people who will be affected by the project will have an influence and early participation of key end users will support the identification of the real issues and the real benefits
		LOW	HIGH
		Stakeholder influence	

Figure 3-11 Stakeholder management

Benefits realization

Defining the benefits criteria is the first stage in the benefits lifecycle, which only ends when the project has been delivered and the benefits have been realized in a sustainable way.

Sustainability should be integrated into any performance management system, and it is usual for benefit metrics to be integrated into operational measures systems to ensure benefits tracking continues long after completion of the project. Sustainability starts with the benefits criteria and should then be cascaded into each aspect of the project.

Benefits concept

Handy hints

Don't forget about the simple set of tools and techniques which are available to support identifying the benefits criteria and then applying them

Project Management Toolkit (Melton, 2007) introduced a selection of tools and techniques which support identification of benefits, benefits criteria and the alignment of benefits criteria to strategic organizational goals – use them!

- The 'Why?' Checklist.
- The Benefits Hierarchy.
- Benefits mapping.
- The Benefits Specification Table.
- Stakeholder mapping.

In linking your project to benefits criteria you also need to consider the overall sponsorship issues

The sponsor can be considered the human link in the relationship between the project and the business. If appropriately chosen, a sponsor can support identification of the appropriate strategic goal and can be a doorway to much of the information needed to identify benefits criteria, thus supporting definition of the project scope.

Not all projects will have a financial benefit. Consider the non-financial benefits which will add value to your business

All organizations need to have a balance between financial and non-financial performance/objectives. It is then likely that some projects will support the achievement of benefit criteria which are linked only to non-financial organizational goals. This does not mean that a robust business case cannot be developed, however it may receive a greater challenge as it will sit outside of the usual economic evaluation processes and rules.

And finally...

- Benefits mapping is an excellent methodology for developing a deeper understanding of your strategic organizational objectives. It helps to strip away the management jargon and put the objectives into an operational frame of reference.
- If there is no link between the potential project (or idea) and a strategic organizational goal then there is no robust business case.

4 Benefits specification – Part 1: linking scope to benefits

Deciding on an appropriate project scope is usually difficult, frequently reveals differences of opinion and occasionally leads to an open conflict. The process can be made much simpler and less stressful by focusing on the intended benefits and their enablers. The project scope needs to be seen in terms of enabling the benefits of the investment to be realized. This linking is the point at which the project becomes defined and provides the input to the development of the business case (Chapter 5).

Defining project scope

At the scoping stage it is very easy to subordinate the business objectives (why the project is being done) to the project objectives (how it will be delivered). For example, the objective of 'streamlining warehouse operations to reduce operating costs' can becomes the objective' installation of a warehouse management system on time and within budget. Whilst this may represent the 'right' project to achieve the business benefits the link between objectives and benefits is often lost and not communicated to all stakeholders.

Traditionally, project objectives have been set in terms of:

- Capital cost.
- Programme or beneficial operation date.
- Functional achievement (based on some form of user requirement specification).

Ideally, these should be compatible with each other and be challenging but achievable. Often Project Teams focus on these objectives or constraints and do not feel that they can be achieved either individually or simultaneously. For example, focus on delivery date can mean the inappropriate removal of deliverables without thought of the consequences for the initial benefit case.

The hierarchy of objectives (Figure 4-1 and Table 4-1) provides a useful framework to explain the sources of project targets and constraints. It is often useful to take Project Teams through this hierarchy and a benefits mapping session (page 31) so that the rationale for the agreed project scope can be understood and bought into.

The hierarchy of objectives can be considered a collation of Benefit Hierarchies (Melton, 2007 and Appendix 12.6) linking the benefits enablers of a number of projects to the corporate vision. It is another tool to link activities to organizational vision and can also be compared to benefits mapping and the critical success factor (CSF) methodology (Melton, 2007). The importance here (no matter the specific tool used) is to ensure that project objectives are linked to the scope that is actually needed; scope that is critical to the delivery of project success and the realization of specified business benefits.

Project Benefits Management

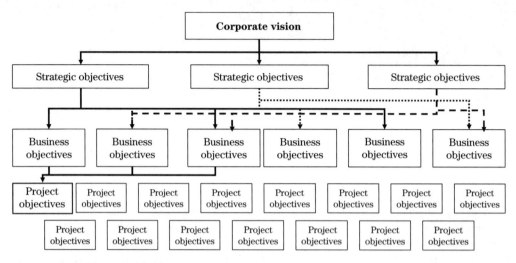

Figure 4-1 The hierarchy of objectives

Table 4-1 The hierarchy of objectives

Level	Explanation	Examples
Corporate vision	What the organization sets itself as a target	To become the largest and most profitable global manufacturer of a specific product by year 5 **Note** Specific targets will need to be set and details added for this to be meaningful as it is passed down to businesses
Strategic objectives	What the organization as a whole needs to do to achieve the vision outlined above	1. Increase market share by X% 2. Reduce net production costs by Y% 3. Expand production into nominated countries 4. Develop marketing partnerships where currently unrepresented 5. Maintain rate of introduction of new products 6. Reduce dependency on supplier of major raw material 7. Maintain a minimum rate of return on investments of 20%
Business objectives	What individual business units need to achieve to enable the organization as a whole to meet its objectives (benefit metrics)	1. Increase production capacity for products A, B and C by 20% 2. Increase yield on product D 3. Improve the quality of products B, D and E 4. Reduce energy consumption on all projects 5. Begin manufacture of raw material Z 6. Maintain a minimum rate of return on investments of 20%
Project objectives	What particular projects need to achieve to enable their business unit to achieve its objectives (benefit enablers)	1. Design and build a production plant for raw material Z 2. Ensure that quality of finished material is no less than 99.95% 3. Upgrade production plants for B and E to meet revised quality standards (partially achieved by project objective 1) and increase capacity by 20% 4. For a total expenditure of no more than €35 million by end of next year

Tool: CTQ Scope Definition Tool

CTQ stands for 'critical to quality' and in the context of the tool a CTQ can be considered a critical feature of the scope which is necessary to meet customer needs (that are assumed to align with business needs). It is based on a Kano Analysis as applied to project scope definition (Melton, 2008) and looks at the critical features of the project scope. A Kano analysis would usually plot critical features required by a customer when receiving a product or service. In a project scope definition context, areas of scope are plotted against business satisfaction with the aim of only allowing it if it delivered a critical feature of the benefits case.

The CTQ Scope Definition Tool (Table 4-2) can therefore be used to both define and challenge scope and would usually be completed by Project Team members. As this would be a different group than those who originally developed the business case it can also provide a challenge to the original Benefits Hierarchy.

Table 4-2 The CTQ Scope Definition Tool explained

Benefits Management Toolkit – CTQ Scope Definition Tool			
Project: <insert project name>		**Project Manager:** <insert name>	
Date: <insert date>		**Project Sponsor:** <insert name>	
Benefit criteria	**Critical feature**	**Project scope**	**Deliverable**
<insert organizational or project benefit criteria>	<insert a CTQ>	<insert scope required to enable the CTQ>	<insert the tangible output from the scope>

Benefit criteria

A benefit area or category which is of interest to the project and the organization. For example, a common benefit criteria is to 'reduce costs'.

Critical feature

A CTQ, or critical feature, is a feature which the customer uses to evaluate the quality of the project outcome. It can be considered a scope feature which enables the business to assess whether the project is likely to deliver the required benefits. For example, the business may have determined that a critical feature of a project to 'reduce operating costs' is a 'reduction in operational personnel'. If the business doesn't see this they won't see the project as being able to deliver the benefit – reduced costs. It is important that the defined CTQs do really define scope criticality rather than just customer desires.

Project scope

High level scope statements should be inserted here and link to the CTQ. At this stage the CTQ or benefit criteria may be challenged if, after brainstorming, the team is unable to find a robust scope which will deliver a specific CTQ or link clearly to a benefit criteria. For example, in order to 'reduce operating personnel' at a specific facility the scope which could deliver this could be facility automation, operational improvements or capacity reduction. In order to select the right scope the benefit criteria needs better definition – 'reduce costs for the same level of output', or the CTQ needs to be changed.

Deliverable

This is a tangible output from the project as a result of delivering the scope. In the example the operating improvements might deliver new operating procedures or a revised layout. This example is detailed in full in Table 4-3.

Table 4-3 Example CTQ Scope Definition Tool

	*Benefits Management Toolkit* – CTQ Scope Definition Tool		
Project: Operation improvement		**Project Manager:** Peter Piper	
Date: Week 1		**Project Sponsor:** John Smith	
Benefit criteria	**Critical feature**	**Project scope**	**Deliverable**
Reduce operating costs per unit output	Reduce number of operating personnel	Improve the process cycle time	• Cell based operational layout • New standard operating procedure (SOP)
	Decrease waste from the operation	Improve the robustness of the process	• New equipment item A • New SOP
	Decrease maintenance costs	Unable to define at this stage	Unable to define at this stage

At this early stage in a project it is always possible that scope can't be fully articulated for all critical features as too much design work would be needed. Different organizations have different ways to overcome this issue; most typically organizations release funds based on early definition specifically to develop the scope.

Challenging the project scope

Although it is desirable that project scope only contains features which meet business objectives and deliver benefits, it is inevitable that other requirements or objectives become wrapped up in the project. For example, there may be additional objectives placed on the project around the use of specific resources or implicit constraints around the use of company procedures or preferred suppliers.

An important part of the development of the project scope is challenging it, to be absolutely clear what is necessary and what is optional. Defining what is necessary requires careful thought as it is easy to persuade oneself that interesting, fashionable or commonly supplied features are 'necessary'.

Scope-benefit analysis

The first level of challenge is to again check that the scope will enable the business benefits to be delivered. If the scope has been developed with this in mind then it should be robust and align with organizational goals as well as support the approved business case.

Benefits specification – Part 1: linking scope to benefits

Cost-benefit analysis

In helping to decide what is required for a project, a useful approach is to adopt a 'triage' methodology. Suggested features are divided into those which are definitely required by the project objectives and accepted, those which are definitely outside the project objectives and are rejected and a smaller third group about which it is not possible to be sure. This remaining group is subjected to greater scrutiny and the process is repeated. This approach has the major advantage that the debate can be structured and decisions can be made final very quickly.

Figure 4-2 shows the triage process as a flow chart. It should be self-evident that when the project scope is agreed, any proposed feature must either be in the scope or outside it. If it is in the scope it is accepted. If it is out of the scope it is rejected. If on the other hand, it can't be decided if it is in or out, then the selection criteria are inadequate and they need to be revised or clarified. This process is repeated until there is a defined scope.

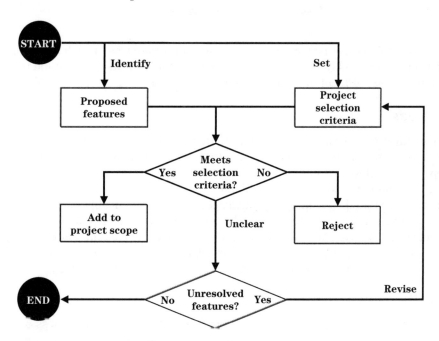

Figure 4-2 The triage method

Having completed a triage based review of the project it is often beneficial to examine the cost-benefit of each feature, using a consistent basis to assess whether value for money is being obtained. The matrix shown in Figure 4-3 presents appropriate actions for each circumstance. If a lean process has been used to develop the project then this stage should be simple.

Organizations tend to be good at assessing the cost of assets but frequently underestimate the benefits. This can explain why cost cutting is so frequent in the project development process. If a cost-benefit approach is adopted, then it is much less likely that a simple cost cutting exercise will be required. Using this approach, features where the cost outweighs the benefit can be rejected, whilst those where the benefit exceeds the cost will be retained. It is worth repeating that if the project objectives are not fully aligned with corporate objectives then the project is less likely to provide value for money. Approaches for deciding on the benefits attributable to the overall project were outlined in

Project Benefits Management

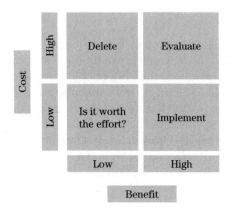

Figure 4-3 Cost-benefit analysis

Chapter 3. Having completed both an assessment of the appropriateness of each item using the triage approach and a cost-benefit analysis, the project can be evaluated for its contribution to the business (providing an effective assessment of the benefits has been made).

Challenging needs

When developing any project scope it is important to distinguish between what the customer or sponsor says he or she wants and what they really need. In most cases, problem owners present their view of what an appropriate solution would be. Most people are able only to articulate their problems by describing what they think the solution will look like.

It is entirely possible that the solution they envisage will be impracticable or inappropriate, particularly if their expertise is in other areas of the business. The challenge in developing an effective solution (project scope) is to get back to the original problem. If the customer is given what they say they want and it is not what they need, the end result might be failure, or worse a white elephant. The needs–wants matrix (Figure 4-4) provides a framework for resolving this difficulty.

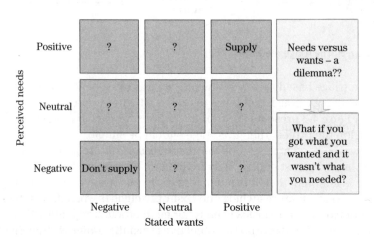

Figure 4-4 The needs–wants matrix

In using this matrix, it should be noted that the cells with the question marks imply a need for further investigation. In doing so, it is crucial that an understanding of the customer's perspective is investigated to ensure that there is no misunderstanding and that views are not based on an incorrect interpretation of the situation. If after further investigation, it is found that the problem owner's view is incorrect, some gentle education will need to be undertaken if a good working relationship is to be maintained! This form of contracting (Melton, 2008) is another aspect of good stakeholder management.

Tool: Scope Challenge Checklist

The Scope Challenge Checklist provides a means to challenge individual elements of scope against criteria that are typical of the type of discretionary scope seen in many projects (Table 4-4).

Any expenditure that does not meet the criteria for 'necessary' is consequently discretionary (i.e. unrelated to the delivery of a project CSF) and should be treated as such. In most cases, this means the scope should be eliminated.

Table 4-4 Scope Challenge Checklist explained

Benefits Management Toolkit – Scope Challenge Checklist			
Project:	<insert project name>	**Project Manager:**	<insert name>
Date:	<insert date>	**Page**	<insert page x of y>
Scope item		**Benefit area**	**Rationale**
<scope item 1>		<impacted benefit area>	<what is the rationale for the scope being included>

Scope item

This is a list of each component of the scope, at a level suitable for the analysis. For example, 'plant automation' would be too high a level, whereas 'new flow control valve in line 123' would be too low. A more logical scope item would be 'full sequence control of the process'.

Benefit area

What benefit can be attributed to this item of scope?

Rationale

Why is this scope item included? The simple expedient of 'needed to achieve the benefit' should not be used here, but a much clearer understanding of where the scope comes from. Typical examples of the types of rationale are:

Forward investment

Many projects provide opportunities to include development activities which generate only small immediate benefits but which would make future projects easier or more productive. It is right for managers to point out these opportunities, but they should not be articulated as being part of the required project scope. After all the future investments may never be implemented and the money may be better spent elsewhere.

Infrastructure

Similar comments apply to infrastructure developments. In most cases, these do not bear economic scrutiny alone and are usually appended to the project that makes their development essential. This can have a distorting effect where a less valuable project (which can be serviced by existing facilities) is easier to justify than the next, more valuable project, which needs the new infrastructure.

Infrastructure expenditure should be clearly identified and highlighted as a separate cost which may ultimately be apportioned to all facilities that make use of it.

Standards

All companies use standards to support design, operation, maintenance and other business processes. The danger here is that blindly applying standards has a cost. Indeed standards always imply additional costs since they reduce the designer's degrees of freedom. The use of standards should be challenged from the perspective of whether they are:

- *Appropriate* – for example, should standards for external colour be issued for the procurement of commercially available equipment that cannot meet the standard without creating a one-off special?
- *Necessary* – for example, should a specification for computer hardware apply to a provider of specialist chromatography equipment who normally uses a different, but technically equal, model?

Functional objectives

Functional specialists may be following their own agendas, normally in the honest belief that this is in the corporate interest. Since most specialists like the challenge of additional functionality, this is a key area where unnecessary cost can be built in. For example, the automation system may be procured with an ability to handle sequence control, but the project does not require that feature now.

Personal objectives

Individuals may have special interests, beliefs or habitual methods that influence what they recommend. They are likely to believe that these approaches are in the company's best interests, but there may be no mechanism in place to validate their opinion. For example, an engineer may prefer to design a system to operate in a very specific way based on previous experiences of always operating in that way. The only way to challenge personal objectives is to be subjective and data rational.

Unclear focus

The most significant cause of unplanned discretionary expenditure is the failure to close the gaps in understanding between various interested parties on the purpose, approach and scope of the project. As noted, projects are strategic initiatives to close the gap between current capabilities and future demands.

Using the Scope Challenge Checklist in a brainstorming session can elicit useful insight into the reason for the inclusion of certain scope items. The example in Table 4-5 illustrates this for the provision of a Manufacturing Execution System in a chemical manufacturing facility.

Table 4-5 Example Scope Challenge Checklist

\multicolumn{4}{c}{**Benefits Management Toolkit – Scope Challenge Checklist**}
Project: MES		**Project Manager:**	Bob Smith
Date: January		**Page**	1 of 1
Scope item	**Benefit area**		**Rationale**
Data historian	Process analysis and understanding		Provides the core collection of data and underpins other functional systems
Site network cabling	Underpins all areas		Infrastructure to support this project and future developments

In this example, the site network cabling is difficult to tie to a direct benefit area and is both an infrastructure issue and a future enabler for the business. Clearly understanding this means that the detailed costs and scope for this element of the project can be reviewed and the justification of the infrastructure element taken outside the actual project.

Short case study

Situation

An engineer was asked by a building manager to investigate a plume rising from the outlet of his plant. He had had several complaints about 'wasting energy' and the plume was visible from most of the site. He suggested that a vent condenser would do the trick.

On investigation it transpired that the building was a very large assembly shop and warehouse and was heated by air handling units using low pressure steam. It turned out that the plume was flash steam generated when the condensate from the system was let down to atmospheric pressure for discharge to drain.

The engineer discovered that there was a return condensate main but it ran at too high a pressure to accept the condensate from the air handling units. He also calculated that the hot condensate could easily be used to heat the incoming cold air in the first bank of heaters.

Potential project

It soon became clear that it would be very cost effective to re-route the condensate through some of the heating coils with a consequent saving in steam. It was also relatively easy to recover the condensate by installing a collection vessel and pump. This would solve the plume problem, save on steam and recover some condensate.

The cost of the piping modifications, the new equipment, instrumentation and controls was modest and the savings substantial. The project was so cost effective that it would have paid back its investment in one winter season. It seemed obvious that the right thing to do was to go ahead with the project.

The outcome

The site was large and had many buildings on it. Buildings of this type were charged an annual fee for steam based on their area, so improving the performance had no benefits for the building manager.

On the other hand, the utilities manager had no budget for work inside production buildings. So the manager with the money would reap no benefit and the manager who would benefit had no money. In this case the final decision was within the authority of the building manager.

As the capital cost associated with the vent condenser was lower than that of the condensate recovery system, it was installed and the plume was removed. However, the company now had an increased utility cost as more cooling water was needed and it had missed the opportunity for an overall saving.

The case study shows how implicit constraints (overhead allocation and transfer pricing issues) can get in the way of project justification. What was the actual organizational goal in this case? Certainly the selected project required capital funding and resulted in increased operating costs – was it a success?

This illustrates the need to get back to the organizational goal and to manage stakeholders in an appropriate way to achieve that goal. The process and rules may need to be challenged to get to a solution which suits the whole organization. In developing the scope (and ultimately the business case) boundaries need to be drawn at the appropriate level and a pitch should be made to the person who has the problem, the money, and who will reap the benefit. This may not be the person who initially raises the problem.

Having defined and challenged the scope, an important aspect is how that scope will be delivered – the project strategy.

Scope and project strategy

Although projects follow a common overall methodology, there is a need to consider the particular circumstances and scope when deciding on a strategy for implementing a project. It is best to have an approach which allows the planning of project specifics to be carried out in a structured manner. Every project requires a specific Project Delivery Plan (Melton, 2008), and the high level aspects need to be defined during early scope development and business case definition.

The decision on an appropriate strategy needs to be taken in the context of other major project choices. In most situations, it is desirable to adopt an iterative approach as project scope, cost, programme and strategy all impact on each other and cannot be set independently.

A useful framework for developing a project strategy is this extract from Rudyard Kipling's 'Six Honest Serving Men' (Kipling, 1902):

> '*I keep six honest serving-men*
> *(they taught me all I knew)*
> *Their names are what and why and when*
> *And how and where and who*'

This can easily be turned into a checklist to ensure that all the appropriate decisions have been made, and is then useful as a means of informing everyone involved. Table 4-6 suggests some questions which may be useful under each heading. Further questions can be added as progression is made through successive projects. It should be appreciated that the 'why?' questions refer back to the objectives.

The simple checklist in Table 4-6 can be supplemented with the How? Checklist (Melton, 2007 and Appendix 12-4). This tool goes into more detail on the essential elements of a Project Delivery Plan (PDP) recognizing that a robust business case not only contains the definition of the benefit enablers (the scope) but also how these are to be delivered.

Table 4-6 Project strategy – questions to ask

Root question	Sample questions
What?	What is the scope? What constraints do we have? What are the key objectives?
Why?	Why are we doing the project? Why now? Why here?
When?	When can we start and when do we have to finish? When will the site be available?
How?	How are we going to do the project (project strategy)? How will we know if we have done a good job?
Where?	Where will the project be built? Where will the main equipment be sourced from?
Who?	Who is our customer? Who else is interested? Who are the key decision makers?

A clear, well articulated and communicated strategy will enable the delivery of benefits, because everyone will understand and be working towards the same objectives. They will also help define and clarify the change in the business that the project will bring about.

Scope and business change management

Most projects require an associated (and often neglected) organizational change process focused on issues such as:

- Management restructuring.
- Changes in work practices.
- Changes in operating procedures.
- New work patterns.
- Skills development.
- Organizational development.

If these 'softer' issues are not properly managed then there is likely to be resistance to the necessary changes and the full benefits of the project may never be seen.

The various participants and stakeholders may also have widely different views of what constitutes success. In practice, it may not be possible to satisfy all of them simultaneously, indeed, some may be mutually exclusive. The best one can hope for is a situation where everyone involved is reasonably satisfied, a process known as 'satisficing'.

Ideally a critical path of success (Melton, 2007) would be developed and be a key tool to use in managing all stakeholders. A critical path of success is made up of a selection of CSFs which deliver the vision of success for the project. Rather than aiming for stakeholders to be reasonably satisfied, the methodology aims to manage the expectations of each stakeholder by gaining buy-in to a set of CSFs (or project scope elements) without which the required business benefits cannot be realized.

A final complication is that the outcomes of some measures may not be apparent for some time after the project is complete. In many ways, the more strategic the objective, the less likely it is that the outcome will become apparent quickly. On a more positive note, there may be beneficial by-products for the organization in terms of competence development, organizational learning and knowledge generation which may also be difficult to assess, particularly if no overt processes for fostering and monitoring their development are in place.

If capital projects have a chequered history in terms of success, business change projects are even less likely to be successful. There are many conflicting views, but most authors agree that in excess of two-thirds of such initiatives fail to meet their objectives and in many cases, there can be no trace of the initiative after a couple of years.

At one level, this is understandable, the objectives are more difficult to quantify and the behaviour of people is harder to predict. This, however, is no excuse for failing to manage the process. Since the behaviour of those involved in the development and use of the asset created is crucial to the success of the project, it is essential that the associated business change initiatives are managed at least as well as the project itself.

Business change processes are often hampered by a degree of resistance from those affected by the change. The causes can be put down to (Kotter and Schlesinger, 1979):

- Individual or group self-interest, frequently based on fear of losing:
  - Salary.
  - Status.
  - Security.
- Different assessments of either the problem or the proposed solution.
- Lack of trust.
- Misunderstanding (for example of motivations).
- Lack of tolerance for change.

It should be clear from this that the relationship between Project Team and user is a major factor in the effectiveness of change programmes. A clear explanation of the proposed changes and the logic behind them will go some way towards avoiding these problems.

Other ways of looking at the reasons for resistance are:

- Loss of control:
  - Imposed change is threatening.
  - Feelings of powerlessness and stress.
- Loss of security:
  - Out of comfort zone.
  - Less support.
  - Questioning own competence or feeling that it is under question.
- Loss of 'face':
  - Implication of failure of previous systems.
  - Feeling of inadequacy.

Typical forms of resistance include:

- Delay.
- Failure to provide information.
- Failure to attend meetings.
- Lack of interest in training.

- Pedantic adherence to policy, procedure or standards.
- Argument with proposals.
- Persistent re-introduction of issues.
- Driving discussions off subject.
- Introducing inappropriate but difficult to deflect/answer questions.
- Repeated failure to comply with changed procedures or processes.
- Negative comments on the company 'grapevine'.
- In extreme cases, sabotage.

Jick (1991) suggested that the reasons why change processes don't work are:

- Implementation takes too long.
- Unexpected problems surface.
- Ineffective co-ordination.
- Competing activities and crises distracted leaders.
- Employee capabilities insufficient.
- Inadequate training at lower levels.
- External environmental factors.
- Inadequate leadership and direction.
- Lack of detailed definition of activities.
- Inadequate information systems.

It should be evident that the use of an effective project methodology should counteract many of these concerns. It is important to understand that the CSFs that will realize the project scope (benefit enablers) need to include a response to some or all of the soft issues raised here (for example, familiarization and training workshops, stakeholder management at the user level, key user sponsors).

Kotter and Schlesinger (1979) present a range of strategies for implementing organizational change and suggest that where possible, the softer approaches based on education, involvement and participation should be used (Figure 4-5). They recognize however that there are some change initiatives which require a more aggressive approach and that there may even be situations where a coercive approach is justified.

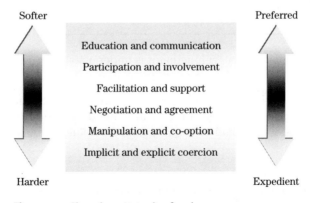

Figure 4-5 Choosing strategies for change

The change management checklist provides a useful vehicle to review the change management aspects of the project. These ideas are similar to the recommendations for effective project management, so apply a similar level of diligence, commitment and effort to changes in business processes and people, as to the project. Collectively a clear, robust scope and an understanding of the impact on the business are the necessary steps towards building the business case for the project.

Tool: Business Environment Checklist

The aim of the Business Environment Checklist is to ensure that all business change issues impacting early scope definition have been reviewed (Table 4-7).

Table 4-7 Business Environment Checklist explained

Benefits Management Toolkit – Business Environment Checklist

Project:	<insert project title>	Project Manager:	<insert name>
Date:	<insert date>	Page	<insert page x of y>

Checklist item	Comment
1. Is a champion supported by a dedicated team, in place and ready to drive the change?	<insert response>
2. Is the purpose and direction of the change clearly articulated?	<insert response>
3. Are the change drivers understood?	<insert response>
4. Is there agreement from key stakeholders on the method and approach?	<insert response>
5. Have you communicated the logic behind the change clearly and concisely?	<insert response>
6. Is there mutual trust and understanding?	<insert response>
7. Are there communication mechanisms and a means to engage in open dialogue?	<insert response>
8. Are people clear about their responsibilities for the change?	<insert response>
9. Is the organization effectively organized to deliver the change?	<insert response>
10. Do the plans envisage quick wins to help build consensus and maintain progress?	<insert response>
11. How will those who lose or are damaged by the change be supported?	<insert response>
12. How will you adapt to changes?	<insert response>

Checklist item

These are typical checks which should be performed to ensure that the business is as ready for the change as is needed at this stage in the project life. Bear in mind that prior to business approval of a project it is unlikely to be understood and/or communicated outside of management teams.

Comment

These should be responses which either indicate that all business change activities are appropriately advanced or alternatively that action is required in specific areas.

Handy hints

If you don't know what you're delivering then you're unlikely to deliver it

Getting the scope right is crucial. If you set out to do the wrong thing then you will not enable the benefits that you and your stakeholders are expecting.

Projects are about people – no participation equals no commitment

The definition of the benefits enablers (project scope) must involve all project stakeholders if risk is to be minimized. Stakeholder management is a project management process which supports project success.

Stand back and look at the 'big picture' – get that right and the details will follow

In scope development, as with many other aspects of project management, it is important to understand the overall aim of the project – the vision of success – and then gradually build up the detail, CSF by CSF, level by level.

Don't expect to get the scope 'right first time'

Scope development and definition is an iterative process. You need to go round the loop a few times. Each loop gives you a better understanding of the problem and the solution, and therefore the appropriate benefit enablers (project scope).

Remember the law of diminishing returns

Although there is no limit to how long you could spend on scope development remember that each iteration will generate less information and deliver less 'value' within the business case development process. At some point you have to decide on what you are going with and you may have to make a judgement call.

Don't forget that projects are a means to an end – focus on the destination and the journey will become clear

The benefit enablers (project scope) must enable the delivery of a set of benefits required by the organization. No project is an island and therefore no project scope can be separate from the needs of the organization.

And finally . . .

- Remember that 'a stitch in time saves nine'. Effort in project scope definition pays dividends later, so don't skimp on the definition.

5 Benefits specification – Part 2: business case development

Much of the development of benefits criteria is focused around the business case. This is the key opportunity to establish why the project should go ahead. Like many things in a project, the business case follows a development cycle as the project evolves from idea to formal proposal (Figure 5-1).

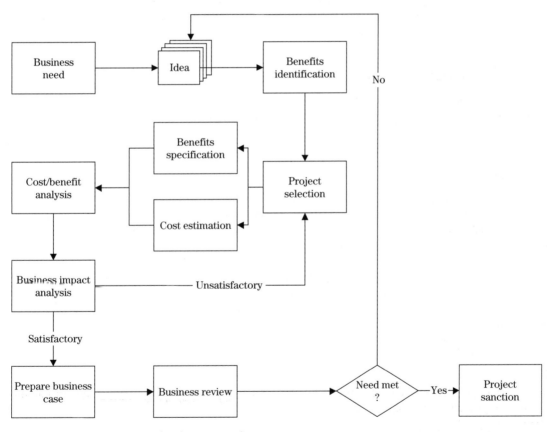

Figure 5-1 The business case development cycle

At key points in this cycle (benefits identification, benefits specification and business impact analysis) there are specific challenges to the choice of project and the deliverables within it. Remember that your project is unlikely to be the only one that the business is considering, and that there is, in effect, a competition for funds. The business case is therefore a sales exercise to persuade senior management to buy a product. If thought of in these terms, it is easy to understand why business cases fail to get approved:

- **The business need hasn't been understood.** Rarely does a business need a new pump or a room full of computers. What a business needs is additional capacity to support enhanced production or a reduction in the cost of processing sales invoices.
- **The business's requirement for return hasn't been met.** In effect, this is the crucial aspect of the business case. If the return is 2.56 years, and the company requires 18 months, it will not be approved (Chapter 2 illustrates other methods of economic evaluation).
- **It has been pitched to the wrong people.** Who will approve the business case? A technical manager, the finance director, or Chief Executive Officer? The audience must be known before they can be sold to!

This can be summarized by using a familiar sales analogy, *'know the audience, know what they want and know how much they have to spend'*. The business case is more formal than just a sales pitch. It is a contract between the project and the business. In return for a given amount of time and money, a promise is made to generate a given benefit for the business.

Value-add and lean thinking

The business case is intended to show the value-add that the project will bring to the business. Lean thinking at this stage of the project is critical to maximizing value and is a useful approach to the cost-benefit analysis.

Lean thinking

In traditional applications of lean thinking, the objective is to eliminate waste (such as rework) in order to maximize the value of a process. Lean thinking is customer focused. It looks at what the customer wants to establish – the 'pull'. The 'value stream' is the mechanism used to deliver the customer requirement. In a production facility (lean concepts originated in the automotive industry), lean looks at the processing stages and the activities between them.

For a project, the pull is the delivery of the business benefits, and the value stream is the project scope (the benefit enablers). Applying lean thinking to development of the business case is geared to achieving the same objective. Maximize those things that are value-add and eliminate (or reduce) those things that are not.

Value

Value as a concept relates to a level of performance for a given resource. The higher the performance, the higher the perceived value.

- Value = performance/resource

Unfortunately, the different stakeholders, users and participants (both external and internal) in a project tend to hold differing views on what represents value. Value management is concerned with reconciling those views to achieve a desired performance (benefit) for the use of minimal resources (cost). It is also important to understand that value can be increased by improving performance, rather than simply reducing resource.

Value management and engineering

Value management within a project is an analytical, prospective approach concerned with establishing value through an analysis of benefits, cost and deliverables:

- Value = benefits/cost.
- The benefits are what the customer wants and are achieved through the deliverables.

Value engineering looks at the individual deliverables in order to evaluate options and alternatives to meeting the benefits. It should also be prospective, although in many projects this approach is used to challenge the defined project scope.

Lean value management

Taken together, lean and value engineering can be used to identify and maximize the value-add in a project. One approach of lean value management is a simple three-step technique to highlight areas for consideration (Figure 5-2).

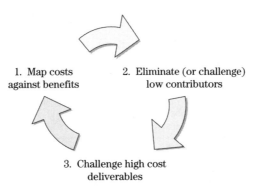

Figure 5-2 Lean value management

This iterative process is a key step in developing the cost-benefit analysis and can extend as the project scope develops. The lean value management approach involves mapping projects against the business benefits to help decide between competing projects (Table 5-1).

Table 5-1 A projects matrix

Benefit	Project			
	Project 1	Project 2	Project 3	Project n
	Cost 1	Cost 2	Cost 3	Cost n
Benefit 1	Impact on the benefit (high/low)	Impact on the benefit (high/low)	Impact on the benefit (high/low)	Impact on the benefit (high/low)
Benefit 2	Impact on the benefit (high/low)	Impact on the benefit (high/low)	Impact on the benefit (high/low)	Impact on the benefit (high/low)
Benefit 3	Impact on the benefit (high/low)	Impact on the benefit (high/low)	Impact on the benefit (high/low)	Impact on the benefit (high/low)
Benefit n	Impact on the benefit (high/low)	Impact on the benefit (high/low)	Impact on the benefit (high/low)	Impact on the benefit (high/low)

As with conventional value mapping, the objective is to see which projects have little impact on the benefits and see if they can be eliminated.

Value management

Value management happens throughout the life of a project, with different issues and approaches being addressed at each stage (Table 5-2).

Table 5-2 Value management in the project lifecycle

Stage	Tools and issues	Outputs
Initiation	• Business and user needs • Benefits sought • Strategic functions • Business options	Project options for consideration
Feasibility and project planning	• Project definition • Verification of business needs • Review of options • Specification of requirements • Procurement strategy	Project definition and PDP
Design	• Value engineering • Cost analysis of functions • Optimize effectiveness of selected design	Detailed design and tender documentation
Implement	• Design and cost review	Refined detailed design implementation
Use	• Project close-out review • Benefits realization	• Lessons learnt • Identifying future needs

Some areas of the value management process are iterative, for example, completion of a feasibility study may require the project basis to be reconsidered.

Within each stage, the objective of the value management exercise (and the tools used) is to challenge the deliverables within the project with a view to maximizing value (not simply minimizing cost). Within the context of business case development, early value management that supports project definition and evaluation is an important process.

Cost estimating

Approach to cost estimating

A robust cost estimate is a critical component of the business case. As a precursor to carrying out the cost estimate, it is essential that the required accuracy is understood. Estimates of greater than ±20–25% are regarded as budgetary and may be used to sanction the next phase of project evaluation. Estimates in the region of ±10% are 'firm' and would normally be used to sanction a project. Developing the cost estimates at different levels of accuracy is usually associated with the stage gate approval process (Chapters 1 and 2). The resources required to develop a cost estimate increase significantly with the need for accuracy.

Typically the cost of carrying out a definitive estimate is between two and four times that of a budget or order of magnitude estimate. As an example, a £5 million capital engineering project will typically cost £50,000–80,000 for a budget cost and £150,000–500,000 for a ±10% definitive cost.

Project cost estimating is usually performed by adding individual project elements into a project total. The number of elements can vary in size and number from a few large chunks of a project with known costs, to hundreds or thousands of discrete tasks or individual work packages. Knowing the level of estimate required will help guide the amount of effort needed to pull it together.

The type of estimation required will depend on the nature of the project. Broadly speaking, they fall into categories of people and materials, and two general approaches, namely top-down and bottom-up:

- Top-down estimates use rule of thumb (based on the last similar project completed) or simple models (such as cost per square meter) to provide an overall project cost. Typically these techniques are used to provide a rapid estimate of the total cost, and provide something of a sanity check on the project scope and scale before too much work has been done.
- Bottom-up estimates are often prepared by suppliers to support their bid process. They involve using a detailed pricing for each work package making up the project. This method is laborious and time consuming, but can result in a fairly accurate estimate (if the work content is well understood). At some point during most project, this level of costing will be required.

For many projects, the bulk of cost is tied to people. In this case, the best way to estimate project cost is to prepare a detailed project schedule and to use the resource management features of that software to identify the number, type and cost of the labour required. Bear in mind that cost estimates that assume staff or asset availabilities or schedule dependencies outside the Project Manager's control should be considered areas of cost risk and managed accordingly.

If equipment is to be acquired, a recent vendor quote will be helpful. However it needs to be borne in mind, that frequently 'similar' cost data is often not similar. Evaluate whether configurations are really similar, what changes are anticipated, and what additional equipment/spares/activities will be required. These same principles apply to using costs from similar projects or service contracts. What the cost data includes and what has been left out needs to be ascertained. It is sometimes useful to take available cost data and apply a complexity factor to get an order of magnitude cost for the project.

Many internal projects tend to be funded from operating expense (departmental budgets, for example). Unless the department is a specific project group, the costs of internal resources tend to be hidden. This is not an excuse for excluding this cost, as it will have two impacts.

- Identified and costed internal resources are much more likely to be made available (the project is paying for people's time).
- The project may not actually be viable.

Costs can usually be estimated with acceptable accuracy by using relevant historical cost data, a well constructed and documented estimating methodology and a good understanding of the work content to be performed. This approach involves putting as much detail into understanding the tasks as possible and generating assumptions with whatever pieces of knowledge may be available. Whether a cost estimate is done well or not, the most likely problems will be:

- Missing scope.
- Not understanding the technical difficulty.
- Change.

Good cost estimating is concerned with understanding the scope, complexity and risk associated with the project deliverables and making an adequate allowance for the unknown. An uncertainty factor may be added to a specific line item in the estimate to cater for this, say 15–30%.

Poor cost estimation is usually apparent through the extensive use of uncertainty factors, large contingencies and very broad cost categories. This type of estimate is little better than a guess. Poor cost estimation will also have a knock-on effect as the project develops, leading to increased change costs to accommodate the missing items (Figure 5-3).

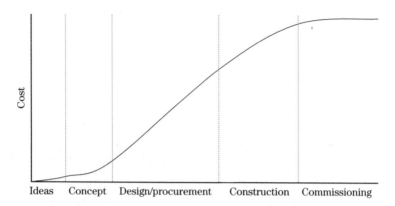

Figure 5-3 Impact of poor cost estimates over a project lifecycle (capital engineering project)

It may be prudent to add a contingency factor to account for expected changes, or to allocate management reserves to deal with later eventualities. Remember however that contingency is 'risk money', not necessarily a formal part of the budget. The contingency should be justifiable (simply adding 10% is not reasonable) and consistent with the risk factors in the project.

Make sure that the cost estimate is well documented and assumptions are recorded thoroughly. If a spreadsheet is being used to prepare the estimate, then all the important data and adjustment factors should be kept visible in cells, rather than hidden in formulas. If the assumptions, factors and data sources are not obvious in the cost estimate documentation, then it is not completed.

A spreadsheet model for cost estimating can also be used to carry out a sensitivity analysis. An uncertain quantitative assumption can be varied and the impact plotted. If the project cost is sensitive to an uncertain assumption, specific efforts should be focused on gathering additional data to reduce that uncertainty.

Financial impact of the project

Although the project will ultimately have a positive benefit on the business (traditionally in terms of a financial return) it will require money to function initially. It is useful to understand where this money will come from and the impact it will have on the business.

Within any organization, there are normally two types of funding available, capital (sometimes known as CAPEX – capital expenditure) and operating expense (sometimes known as OPEX – operating expenditure and sometimes referred to as revenue). There are specific accounting rules about what should be considered capital and operating expense, but broadly speaking the distinction is that the acquisition of assets is capital and the use of those assets is operating expense. In project terms, a project to acquire and install a new pump would be capital funded, whilst a process improvement project using only peoples time would be operating expense funded.

The impact that the different sources of funds have on the business is also important. Capital projects represent a change on the balance sheet (reduction in cash and increase in asset value) and a charge on the profit and loss account related to the annual depreciation of the asset.

For example, a £100,000 project to install new capital equipment has been carried out. The new asset will be depreciated over 10 years according to company accounting policy. The balance sheet will show:

- *Cash* – reduced by £100,000.
- *Assets* – increased by £100,000.

The profit and loss account will show an annual charge of £10,000 for the next 10 years.

If the project were funded out of operating expense, then there would be no impact on the balance sheet, but the profit and loss account would show a charge of £100,000.

Funding projects out of capital therefore has a less immediate impact on the business profitability (assuming cash is on hand), but does have a long-term effect. Bear in mind however, that this is an accounting convention, capital projects still need cash to operate. It is the source of the funds that is different.

Whilst most issues concerning the availability of funding should not impact the Project Manager significantly, there are specific areas he needs to be aware of:

- Access to operating expense funding tends to be time limited. Because the impact on the profit and loss account is immediate, moving costs around due to project slippage over year end can have a very visible impact.
- There is a hidden cost associated with the depreciation charge on a business. As an example, if the £100,000 capital funded project was rendered obsolete and disposed of after 5 years of use, the remaining capital depreciation of the asset would need to be taken as an immediate charge on the profit and loss account (i.e. £50,000 of effective cost). This kind of hidden charge can significantly change the return on a project.

For example, a major pharmaceutical manufacturer was considering a control system upgrade project which would deliver significant benefit. In terms of the cost and the financial return, the project benefit was obvious. However, when reviewing with the business accountant, it was revealed that the book value (un-depreciated cost) of the existing control system was still $300,000, rendering the project financially unviable.

Benefits specification and measurement

The benefit of carrying out a project is not the same as the scope of the project. For example, a new automation system for a plant is not in itself a benefit. Reduced manning, greater consistency of operation, reduced waste and so on are benefits.

It is usual to start (at least at the technical level) with a suggested project scope, launch into the business case (explaining the technical benefits of various acronyms) and then ask for a substantial amount of money. Inevitably such an approach results in a refusal, followed by complaints about how senior management cannot see the logical necessity of the project. It is important to understand that most companies have extremely sophisticated approval systems designed to stop such projects before they see the light of day.

The benefits should be thought through at this stage, as the correct specification of the benefits and their relationship to the scope is the starting point for success. Revisiting the Benefits Hierarchy (Melton, 2007 and Figure 1-9) is a useful place to begin.

The areas of chief concern at this stage are:

- *Benefit criteria* – why are we doing this project?
- *Business case* – what is the cost-benefit analysis?
- *Benefit enablers (project scope)* – what does the project have to deliver to enable the benefits to be realized?

Other project objectives have been covered in Chapter 4 and benefit metrics are covered later in this chapter.

Business need

An additional, external, component to the Benefits Hierarchy is the business need. Is there something that the business is doing (or would like to do) that requires a change? Remember that all projects exist to change something. Knowing the business need is a useful way of framing the overall benefits discussion, and may be an ideal way to pitch the project to senior management.

Identifying the problem statement (or business need) can be tricky (Project Managers tend to like to build, enhance or develop things, but having to shut a plant down due to overcapacity is not necessarily a project to relish). This is where the project sponsor comes in. He or she should have the necessary understanding of the business to help frame this initial need. It should be possible to express the business need in one or two short sentences at most, for example:

- 'To remain competitive in world markets, the manufactured cost for product X needs to be reduced'.
- 'Time lost to IT related problems is too high'.

Benefit criteria

Having the business need clear, the benefit criteria can be identified through an evaluation of the things that influence the business need. In the first example, it is likely that the key influencers are:

- Raw material cost.
- Energy cost.
- Production cost.
- Inventory cost.

In the IT example, they may be:

- Server uptime.
- Desktop query handling.
- User ability.

The benefit criteria that relates to the project can then be clearly identified and stated. This step in the process is a critical one, as picking the wrong set of criteria can help generate the wrong project (for example the raw material cost may represent 85% of the manufactured cost, reducing production cost should really revolve around optimizing procurement).

Support for this approach to mapping business need to criteria can be found in *Project Management Toolkit* (Melton, 2007) and is also summarized in Chapter 3.

Benefits specification

Having established the benefits criteria (the reason why the project is being done), it is then necessary to specify the actual benefits that will be generated. Benefits fall into two main types; hard and soft.

Hard benefits

Hard benefits tend to be those that are tangible, can be (relatively) easily measured and have a cost/value associated with them. Typical examples include:

- Direct labour.
- Indirect labour (support staff/technical, who tend to cost more than direct labour).
- Yield.
- Throughput.
- Asset utilization (as linked to cycle time, downtime and so on).
- Inventory.

At this stage it is extremely important to identify the benefit areas that have significant impact on the overall criteria. There is no value in detailed calculations of the impact of a 50% improvement in an area that only contributes 1% of the total!

Having identified the key benefit areas for specification, the next stage is to calculate the impact and the value of the impact. This is probably the hardest part of any benefits assessment exercise, and needs to be approached as such. Important groundwork items that need to be in place to do this are:

- **Measures of the benefit area that is in need of change:**
  If a measure does not already exist, the benefit area either needs to be discounted, measurement information rapidly obtained, or comparable sources for a measurement examined (with preparation to defend the similarities). A good example is the cost of sales invoice processing as a benefit which feeds into the justification for a new IT project. Does the measurement exist in the department? Can it be calculated from the data held (salary bills and invoices processed) or could industry norms be sourced and applied?
- **Actual value of the change:**
  Even with the measurement information available, how does that apply to the change? For example, how can the benefit of an increase in throughput be measured? Perhaps via a reduction in fixed cost per kilogram, an increase in sales revenue, deferred capital spending (by not building a new plant) or a combination of all three?
- **Practicality of the change:**
  Can the change actually be achieved? For example, a 28% reduction in staffing sounds dramatic, but what is the practical impact. Does it mean redeployment (in which case the cost is moved not eliminated) or are there other factors (for example cover for breaks, training, holiday cover) which make it unlikely to be practical?
- **Who is the owner of the change? Who will be responsible for its delivery?**
  If a 2% yield improvement is being claimed (worth £100,000 per annum) on the project, has the impact on the people who have to live with the change been considered? In this case it may be the Technical Director or Production Manager, who could raise objections to the project unless they are engaged at the benefits specification stage.
- **Is the benefit real?**
  Are the benefits being claimed real benefits or are they an outcome of the project with little or no business impact? A benefit will only be real if, as a result of the change, something can be done. For example, reducing the cycle time for a process will only be beneficial if the released capacity is actually used. Otherwise the plant will stand idle for longer.

Another area to consider, particularly with smaller projects, is that they have a tendency to be blind to things outside the immediate area of the project. For example, a project claiming a reduced

Project Benefits Management

headcount in one area may be unachievable because the people working in that area have other tasks to do which the project will not affect.

A final point to consider in the specification of benefits is their often cumulative nature. By taking each element in turn, and then adding them together, it is possible that the project could be promising to deliver more than is physically possible.

Soft benefits

Soft benefits are those that are intangible, cannot be easily measured and/or are difficult to apply a cost/value to.

Some soft benefits are those things that would be expected of any project (work would not be sanctioned to reduce compliance or decrease customer satisfaction). Nevertheless, soft benefits are an important part of the project justification process. For example, automation reduces manpower (a hard benefit) and through the reduction in human interaction with the process it eliminates errors and therefore improves regulatory compliance (a soft benefit).

When identifying soft benefits, an attempt should be made to break down the benefit area from a nebulous concept (enhance customer satisfaction) to more discrete items (for example on-time delivery and order acknowledgement turnaround time). It may then be possible to add a non-financial measure to these more discrete items. Some business change project may be completely focused around the improvement in these softer benefits and in these case it is imperative that objectives measures are developed.

The same mapping technique that was used for hard benefits can be utilized to develop the relationship between current business processes and soft benefit areas. The current business processes influence the projects ability to achieve soft benefits and this needs to be identified and managed. The influences are clearly the same as for hard benefits and these would be reviewed during specification also. Typically, there would be no more than four to six soft benefit areas explicitly identified as part of the justification for the project on the basis that any more than this and the project is either very large and complex or 'fuzzy' and poorly scoped.

Tool: Benefits Influence Matrix

In order to specify all benefit areas and types, it is necessary to identify what factors are likely to influence them. A useful tool here is to look at the things the project is likely to change or influence (for example manual intervention in a process) and map those to the benefit areas, as in Table 5-3.

Table 5-3 Benefits Influence Matrix explained

Benefits Management Toolkit – Benefits Influence Matrix				
Project:	<insert project name>	**Project Manager:**	<insert name>	
Date:	<insert date>	**Project Sponsor:**	<insert name>	
Benefit area	**Business process impact**			
	Process 1	Process 2	Process 3	Process n
Benefit area 1	<insert level of influence>	<insert level of influence>	<insert level of influence>	<insert level of influence>
Benefit area 2				
Benefit area y				

Benefit area

List the identified benefit areas that impact on the overall organizational goal of the project. These should be at a fairly high level and will be taken from the benefits identified in the initial project scoping exercise. For example, increase in throughput or reduction in manning. These may match with the high level benefits criteria established during benefits mapping.

Business process impact

A project changes the business and therefore changes the business processes within an organization. Therefore all business processes which are likely to influence the benefit areas should be identified. For example, within a Customer Service Review project, how sales orders are handled will impact the identified benefits area of 'on-time delivery'. The business process itself is outside the scope of the project, which is to set up a customer account organization, but how it 'works' will have an impact on the project scope (Table 5-4).

Level of influence

The level of influence of any business process on any benefit area should be ranked as high (a major business change issue), low (a minor business change issue) or none. This approach is to avoid 'fence sitting' with all business processes having a medium influence on all benefit areas. This map will provide a means of identifying where to expend the effort in calculating benefit potential. Any benefit area that has less than one high level of influence should be disregarded in terms of inclusion or consideration within the business change plan, sustainability plan or scope. Any with three or more should be focused on. Those in the middle should be considered on merit and on the available time and effort to review.

Short case study

A production facility needs to enhance its customer service to maintain and grow market share (the product is a commodity item with market pricing). The results of a study to review business process impact are shown in Table 5-4.

Table 5-4 Example Benefits Influence Matrix

Benefits Management Toolkit – Benefits Influence Matrix				
Project: Customer service review		**Project Manager:** confidential		
Date: confidential		**Project Sponsor:** confidential		
Benefit area	**Business process impact**			
	Sales order handling	Production downtime	Stock holding	Technical support group
On-time delivery	High	High	High	Low
Flexibility	High	Low	High	Low
Customer support	Low	Low	High	High

As a result of this review the Project Team integrated a review of the stock holding and sales order handling business processes into their PDP (section on business change). They highlighted that customer service improvements required changes which were outside the current project scope but which needed to occur if the hard and soft benefits were to be sustainably realized.

Benefit Specification Table

Having carried out the mapping and identification of the benefits, the Benefits Specification Table (Appendix 12-8, extracted from *Project Management Toolkit*) provides a record of benefits for use later in the project. Remember, these are benefits that the project will have to deliver.

Benefit measurement

With an established set of benefits for the project (and the business case), now it is a good point to carry out a check that the selected benefits are actually measurable.

Being able to measure the benefits as the project is carried out will make the project credible, but care should be taken to ensure that the measurements are not too complex and do not drive adverse behaviour. For example, measuring the performance of an accounting group by looking at the time taken to process invoices may drive responses such as rejecting invoices or passing activities to other departments, when the intention was to streamline the invoice process. Measurement and behaviour are a particular issue when one department's performance is measured without thinking of the consequences across the business.

When developing measurements, ensure that the following questions can be answered:

- Can it be measured?
- Can it be measured simply (no complex system has to be put in place to develop the measure)?
- Who is directly affected?
- If it was you being measured, what would your response be?
- Is the measure sustainable? Is the measure likely to create a rush of activity to meet a target, followed by a gradual slide back to the original condition?
- Will the measure have a positive impact on behaviour?

How to write a business case

As discussed in the introduction, the business case is both a sales tool for the Project Team to obtain funding, and a contract between the project and the business. A business case can be anything from a single page to more than 20. However it should be possible to explain the key points of the business case in one page, irrespective of the project cost, size or complexity.

The short (one page) business case is ideal as a tool to provide focus for the whole business case development process, and also as a stand-alone document for the approval of projects that have limited impact outside the immediate originators group. Longer business case documents tend to be required by internal processes that specify a particular format and/or justification for the project. They are also useful where the project may have an impact outside the immediate user group or where the reviewers may not have as much understanding of the background. As a general guide, the business case should be as short as possible, bearing in mind its goal.

A one page business case document

The Business Case Tool (Appendix 12-9), extracted from *Project Management Toolkit* provides a succinct summary of the critical components of a business case. It is a useful starting point to develop the detail, as well as providing a discussion tool for working with the project sponsor.

Critical areas that need to be considered are:

- *Background* – why this business area, why this project?
- *Project description* – summary scope and key deliverables.
- *Delivery analysis* – the costs in terms of internal and external resources, capital and revenue.
- *Business change analysis* – impact of the project on the business.
- *Value-add analysis* – the detailed Benefits Hierarchy detailing benefits criteria, cost-benefit and benefits metrics.
- *Impact of NOT doing the project* – what are the alternatives, what is the effect of not following the project now, or ever?

Establishing these areas creates the framework for a typical organizational business case document.

A typical business case document

Typically, the format of the business case needs to conform to a company standard, but the content is generally similar:

- Executive summary.
- Introduction.
- Project scope and organization.
- Benefit and cost.
- Timing.
- Other.

A typical template is included in Appendix 12-5.

Business case fundamentals

Before starting to write a business case, there are a number of fundamentals that need to be considered.

Format

What is the actual format and content required by the organization for a business case? Does this change with value? Preferably a template should be obtained along with an example of a business case that was approved. Make sure this format is adhered to.

Approval process

The approval process for the business case needs to be understood. Questions to ask are:

- What level of management approval is required for what value project?
- Who is actually going to be the approver(s)?
- Will approval have to go off-site? Invariably this means that the approval process can be slowed down significantly.

- What is the approval timescale? Does a management board meet once a month or can approvals be handled when submitted? Factor this into the project planning at an early stage.
- Where does the sponsor/champion fit into the approval process? Bear in mind that the sponsor is ultimately responsible for delivering the promised benefit and should be able to front the business case to the approver(s).

Understanding the approval process will help in understanding the audience.

Writing style

Whilst word processors eliminate the majority of typos on written documents, there is nothing more likely to cause rejection than an incoherent, repetitive or grammatically incorrect business case. This tends to be more important the higher up the approval chain the business case goes.

Write the business case clearly, succinctly (if it can't be explained in simple sentences, it hasn't been thought through) and avoid the more obvious grammatical errors.

Business case content

The sections of a business case that tend to get read first are the executive summary, the benefits and the cost. Within these three sections, it is necessary to capture the attention of the reader, convince them of the suitability of the project and ask them for the resources (funds, assets, people). The rest of the document demonstrates how the project is to be delivered, what the alternatives are and what risk factors exist.

Executive summary

This is the section where there is an opportunity to capture the attention of the reader. The broad objective and benefit of the project should be outlined. It needs to be short and to the point, a half-page made up of two or three paragraphs should suffice.

It may help to think of this in terms of what is sometimes known as the 'elevator pitch'. Imagine you have got into the elevator (lift) with the project approver and you have 30 seconds to explain your project objectives and benefits. It is also useful here to consider the benefits in terms of payback to the business (Chapter 2), rather than explicitly reference the cost and benefit numbers.

Introduction

The introduction provides the problem statement, the background to the project, why it was selected, how the business will be affected and so on. Here it should be demonstrated that the business needs have been considered, a number of options have been identified, and a suitable one selected.

Project scope and organization

The scope of the project should be defined in simple terms. Talk in terms of major deliverables and activities. Avoid diving into too much detail at this stage, as the objective is to provide a taste of the overall project.

Project organization should consider the nature of resources required, both internally and externally, to deliver the project.

Benefits specification – Part 2: business case development

Project delivery strategy

Delivery strategy is concerned primarily with the 'how?' of the project, rather than the 'why?'. Within the business case, the overall approach to delivering the project should be considered. Most elements which make up a formal Project Delivery Plan (PDP) should be considered (Melton, 2008) such as:

- Project roadmap (key stages and stage gates).
- Project controls (cost, scope, time, risk, change).
- Benefits management.
- Engagement of the business.

There should not be any great detail at this stage, however, it should be demonstrated that how the project will be delivered when it is approved has been thought about.

Benefit and cost

The statement of benefits within a business case tends to be somewhat anti-climactic considering the amount of work undertaken to develop them in the first case. The project approvers will, in the first instance, be looking for a simple statement that the benefit is greater than the cost (within some defined business criteria). It is probably easiest to make this benefit/cost statement within the opening sentence of this section, for example: *'The estimated project costs are £150,000 with an anticipated payback of 9 months'*. The same sentence will work for the executive summary.

The remainder of this section should contain a summary of the benefit and cost data that has been collected as part of the specification exercise. It should contain both a summary table of key cost and benefit areas and explain how the estimates have been made (reference to quotations for cost and data assembled for benefits). Don't be afraid to make use of figures to illustrate the cost-benefit analysis (Figure 5-4).

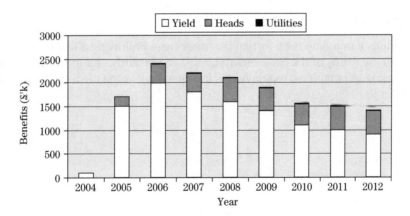

Figure 5-4 Example benefit analysis

The costs and benefits analysis should also consider sensitivity within the calculations. It should include critical sensitivities within the benefits discussion. Most sensitivities at this stage of a project tend to relate to throughput; number of documents processed, market demand and so on. Care needs to be taken to distinguish between sensitivities and risk factors. Sensitivities are variables outside the control of the Project Manager that will impact the return. Risk factors may be controllable, are internal to the project (even if they have external originators) and impact the ability of the project to deliver.

Timing

The timing section should contain a top level plan of the project indicating the key milestones and deliverables.

Others

This section tends to be a catch-all for background on the project, or in some cases, may not be needed at all. Some of the content described in detail here may appear in other sections. Nevertheless, the key issues that should be considered are:

- *Alternatives* – what else could be done or what if no project is undertaken? Remember that doing nothing carries a risk, and this will need to be highlighted in the business case.
- *Risk* – what risks does the project itself carry? Potential impact on production, failure to meet specific deadlines and so on:
  - The risk analysis should identify constraints on the project in terms of budget, schedule, scope and quality. A plan to minimize the impact of these constraints should be identified.
  - Describe how the Project Manager will identify issues or risks during the project.
  - Known risks to the successful completion of the project should be identified and plans to minimize those risks outlined.
  - The risk analysis should answer the following questions:
    - What are the chances that the project will be successful?
    - What will be done to maximize the chances for success?
    - Is there enough contingency in the budget and schedule?

Business case template

Appendix 12-5 contains a template for a formal business case. Following a formal structure ensures that your proposal meets the needs of the business and, most importantly, that information required for key decisions is not missing. A completed example of a business case using the template is contained in the case study in Chapter 9.

Handy hints

Don't forget about the simple set of tools and techniques which are available to support the creation of the business case

Project Management Toolkit (Melton, 2007) introduced a selection of tools and techniques which support identification of benefits, benefits criteria and the alignment of benefits criteria to strategic organizational goals – use them!

- The 'Why?' Checklist.
- The Benefits Hierarchy.
- Benefits mapping.
- The Benefits Specification Table.
- Business Case Tool.

Keep it simple!

You will be asking for funding from busy people. Your business case needs to be written in clear and simple language so that the reader immediately understands why they are being asked to fund. Complexity, jargon and demonstrating technical brilliance have no place in a business case. Remember, if you can't explain the basic purpose of the project in two or three sentences, you need to do more work on it.

Engage your sponsor and/or champion

These are the people who will help guide the development of the business case to a successful conclusion. Make use of their knowledge of the business to produce a relevant business case.

And finally . . .

- Understand what the project will do in terms of responding to a business need.
- Know your business case audience – those who have influence and power over its approval.
- Remember that the business case is both a sales tool and a contract for delivery. It is not a technical project definition. Write it with this in mind.

6 Benefits realization

The project business case defines the expectations the business has: 'why' the business needs the project to be delivered. The Project Delivery Plan (PDP) (Melton, 2008) then details:

- 'What' exact benefits will be delivered (the specification of the benefits).
- 'When' and 'how' benefits will be delivered (benefits realization planning).
- 'What' needs to change within the business in order to support the benefits and how sustainability is measured (business change and sustainability planning).

Benefits realization is at the other extreme of the project lifecycle, but success relies on a robust business case and PDP and then effective project delivery. As defined in *Project Management Toolkit* (Melton, 2007) benefits realization is concerned with the:

- Tracking of benefits delivery after the project scope has been delivered.
- Assessing if the achievement of the project objectives will allow the realization of the improvement in specific benefit metrics.

The benefits realization process (Figure 6-1) cannot be separated from the preceding stages of the benefits management lifecycle (Figure 1-12, page 17). Benefits realization builds on the specification

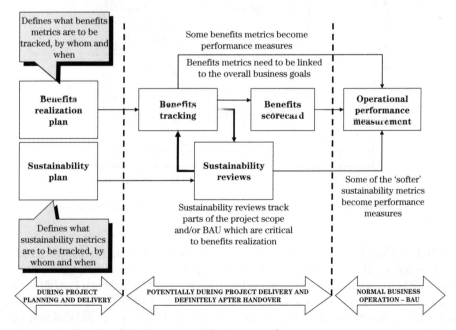

Figure 6-1 Benefits realization process (Melton, 2007)

defined during the earlier stages of the project. It is the point at which the link between the project deliverables which have been completed and the benefits they have enabled are confirmed.

Benefits realization is at the end of the project lifecycle (Figure 1-1, page 1) and it is the one stage where a Project Manager's role may be quite different depending on project type and organizational practices. However, they would be defined in the Benefits Realization Plan (Appendix 12-12) and developed during usual project delivery planning activities. Typically a Project Manager would expect to have a reduction in overall responsibility once project scope has been delivered (Figure 6-2). This profile will change depending on different factors:

- If scope is completed and handed over in phases then the Project Manager must retain a high level of responsibility to ensure that all scope is fully handed over.
- In large construction projects based on fixed prices it is not unusual for the Project Manager to go from 100% to 0% responsibility upon completion, handover and close-out, which occur simultaneously.
- In business change projects it is not unusual for the Project Manager to remain involved with a high level of responsibility until early benefits are realized. Project close-out is then held after this stage.

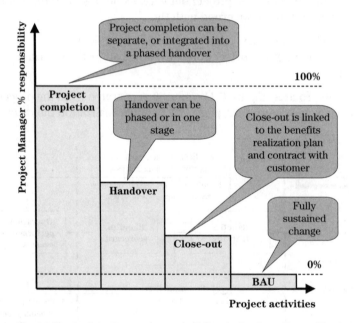

Figure 6-2 Project Manager responsibility during benefits realization

As a Project Manager's responsibility decreases he may disengage from the project, and therefore the sponsor and customer. However, at this stage of disengagement he will still have a high degree of influence which needs to be used appropriately. There is a strong link between benefits realization, business change, customer satisfaction and sustainability. In simple terms, the benefits will not be fully realized if the business change is not implemented in a sustainable manner, and this will not happen if the customer is not satisfied with the project delivery. Therefore the Project Manager needs to act more like a consultant with the customer(s) in order to support this phase of handover, still using his influence.

The consultancy lifecycle (Figure 6-3) follows four key stages, which although generic (Melton, 2008) can also be described in terms of the benefits realization stage of a project.

- **Gaining entry:** Establishes a relationship during the benefits realization phase, gaining a common understanding of the change in relationship from previous phases.
- **Contracting:** Defines the roles of the Project Manager and customer during benefits realization, particularly if elements of the project have, or are shortly due to be, handed over. Agree the level of responsibility and recognize when each other has a strong influencing part to play due to their level of expertise and experience.
- **Engagement:** This is the way that the Project Manager and customer operate as benefits realization progresses, with open and honest communications, maintaining and updating the contract so that the project achieves success.
- **Disengagement:** Establishes agreement that the endpoint has been reached, defines business as usual (BAU) so that both Project Manager and customer manages to 'say goodbye'. This would usually be linked to a final review to demonstrate that the project has been fully and sustainably integrated into the business and is delivering the intended benefits.

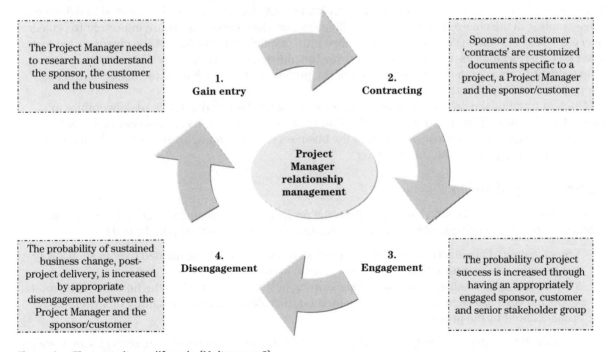

Figure 6-3 The consultancy lifecycle (Melton, 2008)

It is important that the Project Manager works effectively with the sponsor and customer during this stage to support integration of the project into the business, sustainability of all changes and tracking of benefits.

Realization planning

As outlined in Figure 6-1, successful benefits realization starts with robust planning. During the planning stage of a project a Project Manager should already be thinking about the final project stage of benefits delivery. Realization planning is one aspect of the business plan within the PDP and covers:

- Identification of benefit metrics, their baseline and target levels.
- Benefit metrics tracking (who? how? when?).
- The link between the project scope and the benefit metrics.
- The link between business change outside of project scope and the benefit metrics.

Within the PDP there is usually a need for a collation of all benefits definition work completed to date, as well as some additional planning work to ensure that the benefits can be realistically and sustainably realized (Melton, 2008). This is usually documented into some form of benefits realization plan, an example of which is included in Appendix 12-12.

How benefits are realized

Benefits are realized through the delivery of project by both the delivery of hard project scope (a new vessel, a replacement business process) and by the change they introduce into the business. Realization planning needs to cover both the explicit benefits that the project is expected to deliver and those implicit in the change that the project will introduce. Both benefit types are important to project success – projects can be perceived to have failed if the implicit expectations of the customer have not been met, just as much as if they fail to deliver the tangible benefits. Benefits realization planning can help to identify areas of the business, outside of the project scope, which need to change in order to support the delivery of both types of benefit.

Much of traditional delivery planning for a project is focused on getting the hard deliverables in place. Benefits management is about ensuring that the associated benefits are also delivered. Since the realization of the benefit is closely aligned to the business change, an important part of planning for realization is engagement with the customer.

Sustainability planning

As outlined in Figure 6-1, a key deliverable from the project delivery planning stage is the sustainability plan (Melton, 2008) which should incorporate a Sustainability Checklist (Appendix 12-11).

- A Sustainability Plan aims to ensure that the project, once complete, is handed over in such a way that benefits can be realized and all changes remain in place.
- A Sustainability Checklist, usually developed during project delivery when the link between project scope and BAU has been fully defined, is a method to challenge the sustainability of any change brought about by the project.

Sustainability planning can help to identify areas of the business, outside of the project scope, which need to change in order to support the delivery of benefits.

Customer contracts

A PDP would usually highlight the proposed relationship with the customer during the final stage of the project, looking at project completion, handover and then benefits and sustainability tracking.

Tool: Benefits Realization Risk Tool

The aim of Benefits Realization Risk Tool (Table 6-1) is to consider all the potential 'failure modes' – those scenarios which will prevent the full and sustainable realization of the project benefits. The failure modes and effects analysis (FMEA) methodology is appropriate to use in this instance because it reviews failure mode detection. Benefits realization risk analysis is an important element of the planning process. If the risks are not properly understood, the ability to deliver benefits will be significantly affected, as the appropriate mitigating actions will not be progressed at an early enough stage to make an impact on the outcome.

Table 6-1 The Benefits Realization Risk Tool explained

Benefits Management Toolkit – Benefits Realization Risk Tool							
Project:	<insert project name>			**Project Manager:**	<insert name>		
Date:	<insert date>			**Project Sponsor:**	<insert name>		
Benefit	**Probability of not achieving**	**Impact of not achieving**	**Ability to detect failure early**	**Sustainability threat**	**Risk priority number**	**Mitigation plan**	
<insert key benefit characteristic>	<insert score>	<insert score>	<insert score>	<insert score>	<insert calculated score>	<insert mitigation plan>	
Scoring system							
Probability	1 = Low 5 = High	Impact	1 = Low 5 = High	Detection	1 = High 5 = Low	Sustainability	1 = Low 5 = High

Benefit

Each specific benefit of the project should be listed. These may be expressed as a benefit concept or a benefit metric depending on the timing of the risk assessment. All benefits, whether explicit or implicit, should be included. An explicit benefit is one which is defined within the approved business case whereas an implicit benefit is one which is implied through the delivery of the project. For example, an explicit benefit from the introduction of a new operating process may be 'improved efficiency of the operation', whereas the implicit benefit might be 'improved team morale'.

Probability of not achieving

For each benefit there should be an assessment of the likelihood that the benefit will **not** be achieved. In order to assess this aspect failure modes should be considered for each benefit and the potential of that failure mode to occur will be reviewed. A scoring system should be defined and would typically be a 1–5 rating:

- 5 = High; there is a strong probability that the failure mode preventing this benefit from being delivered will occur.
- 1 = Low; there is a small probability that the failure mode preventing this benefit from being delivered will occur.

Impact of not achieving

During the development of the business case there will have been an assessment of the relative importance or weighting of each identified benefit, and therefore the impact of **not** achieving it. A scoring system should be defined and would typically be a 1–5 rating:

- 5 = High; there is a serious impact on the achievement of the approved business case.
- 1 = Low; there is a minimal impact of the achievement of the approved business case.

Ability to detect failure early

It is important to understand how early in the project the failure mode would be detected. This strongly supports forecasting of benefits delivery failure, recognizing that some benefits failure won't be apparent until well after completion and handover. A scoring system should be defined and would typically be a 1–5 rating:

- 5 = Post-completion; the failure of the benefits delivery and any associated failure mode cannot be detected until after project completion and handover.
- 2, 3 or 4 = During project delivery; the failure mode can be detected at some stage during project delivery.
- 1 = Early; the failure mode is built into the planning stage and benefits delivery is easy to forecast based on the planning.

Sustainability threat

It is also important to understand how the benefit can be sustained and the likelihood that the benefit cannot be sustained. A scoring system should be defined and would typically be a 1–5 rating:

- 5 = High; there is a high threat to sustainability.
- 1 = Low; there is a low threat to sustainability.

Risk priority number

A risk priority number can be calculated by multiplying the probability, impact, detection and sustainability rating. A high score indicates a high risk of not delivering the benefit, so should be considered a high priority and an appropriate mitigation plan developed. The individual scores will show where the effort for mitigation needs to be focused. The highest scores should be mitigated as a high priority whilst the lowest scores may have no mitigating actions assigned.

Mitigation plan

It is important to identify what actions need to be undertaken to mitigate the risk to check that the mitigating actions are working the risk assessment should be carried out at regular intervals during project and benefits delivery. If the mitigation plans are effective then the risk priority numbers should decrease. At this stage the focus of mitigation may move to some of the medium risks; recognizing that no organization, or Project Team, has limitless resources to address every risk.

Short case study

A manufacturing organization had approved a project to install an electronic batch record system into its newest manufacturing facility. During project delivery planning the Project Manager conducted a benefits risk assessment in order to check that all required mitigation plans were in place (Table 6-2).

Benefits realization

Table 6-2 Example Benefit Realization Risk Tool

Benefits Management Toolkit – Benefits Realization Risk Tool

Project:	Electronic batch record			Project Manager:		Bob Smith
Date:	April			Project Sponsor:		William King
Benefit	**Probability of not achieving**	**Impact of not achieving**	**Ability to detect failure early**	**Sustainability threat**	**Risk priority number**	**Mitigation plan**
Paperless batch record	3	5	1	1	15	None
Exception based manufacturing review	3	5	3	5	225	Formal review and training required of all teams – engage in design and development process
Exception based quality review	3	5	3	5	225	Formal review and training required of all teams – engage in design and development process

Scoring system

Probability	1 = Low 5 = High	Impact	1 = Low 5 = High	Detection	1 = High 5 = Low	Sustainability	1 = Low 5 = High

As a result of the review the Project Manager identified that the two key benefits required significant mitigation if they were to be achieved. The critical area which required mitigation was the level of sustainability threat, indicating that it was the business change issues which needed to be further reviewed. As a result the Project Manager was able to integrate BAU resources into the team to support the design and testing of the new system. In addition, the customer was able to schedule appropriate review and training sessions to support sustainability once the project was completed and handed over.

Delivery of the explicit benefits

Delivery of the explicit benefits can only occur when the project scope has been delivered. Realizing these benefits proves that the business case has been delivered. Explicit benefits realization takes the benefits specification developed in the planning stage (Appendix 12-8) and uses it as a tracking tool to monitor progress over the life of the project (Appendix 12-10).

Tracking explicit benefit delivery

Benefits are linked to specific areas of scope and/or business changes outside of the project scope. Therefore they can only be delivered once the specific areas of scope are complete or advanced enough for partial benefits to be delivered. In order to adequately track explicit benefits delivery a plan will already have been developed showing:

- The exact benefit metric which demonstrates benefits delivery by achieving predefined targets linked to both benefit metric level and time.
- An appropriate format to track the benefit so that the appropriate action can be taken if the metric deviates from the plan.

The delivery of the explicit benefits should be tracked as these are a part of the contract between the project and the business. Therefore benefits metrics progress measurement is needed to:

- Confirm benefits realization status.
- Assess performance against the benefits realization plan: 'are we going off track?'
- Support forecasting: 'when will all future benefits be delivered?'
- Enable benefits tracking reporting.

The choice of benefit metric will have determined how progress can be measured and therefore how it should be analysed. The key feature of any benefits measures system is that it should support appropriate decision making if the metrics are not achieving planned progress. Typical benefits progress measures systems are:

- *Progress against plan* – using an appropriate unit which can be converted to a percentage complete, or a rating against a score (often used for less tangible benefit metrics) or some form of RAG analysis (where RAG = Red, Amber, Green).
- *Achievement of milestones* – attainment of a tangible benefit target on an agreed date.
- *Earned value* – reviewing benefit target and schedule adherence simultaneously by tracking tangible benefit metrics and the business change/schedule it took to enable them.

Benefits tracking reporting

The aim of any report is to summarize progress so that appropriate actions can be taken. A benefits tracking report is no different although it can take many forms which themselves direct the type of action.

The milestone report

The milestone report is probably one of the simplest forms of benefit report (Figure 6-4). It merely highlights an end status for a benefit metric, although this can also be linked to a specific target

Benefit metric	Milestone target	Current progress
Meet sales demand	Month 2 after project completion	Achieved
Decrease cost to manufacture goods	15% cost reduction upon project completion	Not achieved – at month 3 post-completion the cost reduction was steady at 12%
Reduce reject goods level	Rejects at <5% upon project completion	Not achieved – at month 3 post-completion the reject level was steady at 7.5%

Figure 6-4 Example benefits milestone report

Benefits realization

time-line. It is useful in helping the business to see that benefits have indeed been achieved, but has limited use in forecasting and action planning unless linked with other tracking tools. The example in Figure 6-4 is shows a manufacturing improvement project where three key milestones were linked to external organizational challenges which the project was attempting to meet. The report was generated 3 months after project completion.

For a milestone report to be appropriate there must be some steering team (sponsor, customer, senior business stakeholders) who are able to review the report and decide to take action when a milestone has not been achieved. Often the column for current progress is used to note a forecast if the milestone is not due to be delivered at the time the report is issued. In the case of the example project (Figure 6-4) the cost and reject rate reduction may have been better controlled via some form of progress trending, so that the potential for not achieving the target could have been seen earlier.

The profile report

Often projects are implemented to generate a balanced change within a business. Therefore a change in a profile is needed to support further action as a project is completed and benefits are being realized (Figure 6-5). A radar or histogram format is usually used to show this. These formats allow a review across all the benefits so that decisions can be made to meet a target profile rather than any one specific benefit. This is useful if the benefits are linked in some way. In the example in Figure 6-5

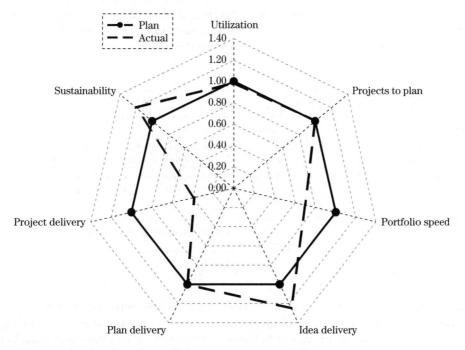

Figure 6-5 Example benefits profile report

a project to improve a business change project group's performance shows the current profile as the project nears completion:

- The team recognize that they need to have a balanced performance in all seven benefits areas.
- The lack of significant improvement in the delivery of projects (this is a quality and speed benefit metric linked to external benchmarks) and the underperformance on portfolio speed are likely to be connected.
- The profile causes the improvement team to review the new standard operating procedure (SOP) for project delivery and to also review associated business change aspects such as selection of appropriate Project Managers (one of the previously identified failure modes for the overall benefits profile).

In this case the benefits profile was an appropriate method to assess benefits delivery and make further adjustments based on the way the profile was developing. The key behind a robust profile report is the method by which each benefit criteria is scored. For instance in Figure 6-5 all benefit metrics were normalized to score out of 125, although each metric originally had different units:

- Utilization was measured as a percentage.
- Portfolio speed was measured as average project time within the portfolio.
- Sustainability was a rating based on percentage of benefits still being realized 6 months after project completion.

The RAG report

For less tangible or measurable benefits a RAG report can be used (Figure 6-6). This report relies on having a target level for each benefit metric which, when achieved, gains a 'green' rating. Forecast achievement or trending towards achievement cannot therefore be given a 'green' rating, this status would usually be an 'amber' rating. The 'red' rating is usually reserved for those benefit metrics where either no progress has been made, or a benefit risk has occurred which will prevent the benefits from being delivered.

Benefit	RAG rating	Comment
Team morale increased	AMBER	There are still a number of team members who are unhappy with the changes
Team more involved with current operations	AMBER	There remain pockets of resistance but over 75% of the team are adapting to the cell working arrangements
Team output has increased	RED	The few resistors are becoming barriers to benefits realization as they are impacting the new lean ways of working
General house keeping has increased	GREEN	The team have taken on board the lean ways of working in terms of keeping the work areas tidy and organized

Figure 6-6 Example benefits RAG report

For additional clarity the RAG rating can be trended so that a benefit metric remaining 'red' over a number of reporting periods is easily identified. In addition a lack of benefit metric sustainability can be seen if a 'green' turns to 'amber' from one report to the next. However, this method of scoring benefits is not very precise and can lead to confusion if the ratings are not used appropriately.

The earned value report

In the context of benefits tracking, earned value has been used to assess if the benefits that are being delivered match their potential based on the enablers that are in place. In other words some benefits should be 'earned' due to project scope and/or business changes which have been completed. In this way the progress report can identify two scenarios:

- *Scenario 1* – actual is less than planned, but earned value matches actual. The benefits are being delivered late but are still forecasted to meet the target levels.
- *Scenario 2* – earned value does not match actual. The benefits are being delivered at a lower level than expected for the scope and business changes delivered and therefore the forecasted target levels are unlikely to be met.

In Figure 6-7 a project to completely renovate a production laboratory shows a benefits tracking trend for financial cost savings linked to the completion of scope: new ways of working (WoW), new equipment and a new laboratory layout.

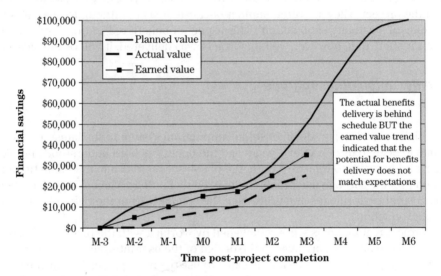

Figure 6-7 Example benefits earned value report

The earned value report at month 3 indicates that:

- The earned value is greater than actual value (financial savings axis) so the benefit enablers (project scope and business changes) are not enabling the expected level of benefits. Therefore the original relationship between enablers and benefits was flawed and some corrective actions need to be investigated.
- The earned value is less than planned value (schedule axis) indicating that the benefit enablers are being delivered slower than planned and the schedule benefits delivery is likely to be delayed.
- The actual is less than planned value (schedule axis) indicating that the benefits are being delivered slower than planned . Without the above two assessments the Project Manager would be unaware that the final benefits outcome would be lower than required as well as late.

Earned value is a powerful process measurement, forecasting and control tool. However it is the set up of the earned value system within the benefits realization plan that is the most crucial element.

Although more complex, the result can support appropriate actions to react quicker to forecast benefits delivery issues.

Mitigating explicit benefit risks

The benefits risk assessments completed during planning and delivery will have already generated a benefits risk profile and associated set of mitigation plans to effectively respond to the risks. This should be used during benefits delivery to identify new risks, monitor current high risks and also watch for any low risks which have now become high. The latter can occur during benefits delivery as sustainability threats become more apparent.

Delivery of the implicit benefits

Delivery of the implicit benefits occurs both during project scope delivery and during formal explicit benefits delivery. Implicit benefits tend to be the softer items involving changes to WoW as a consequence of the project. They are often implicit because they are difficult to quantify or fall under the general heading of 'customer satisfaction'. This aspect of benefit realization is critical. If the customer is not satisfied, then however well the project was delivered, it will be considered a failure.

Short case study: automation upgrade

Situation

A manufacturer undertook to upgrade an obsolete automation system to the latest technology. The existing system had been in use for over 7 years and the operators both understood and liked it. The key project deliverables were to remove a potential obsolescence risk and to enable tighter control of the production process through use of more advanced technology. The business case was based on maintaining current performance and protecting the company from the risk of facility failure (and therefore loss of production).

The detailed project requirements were established by working with the technical personnel on the facility. The operations team were involved in the design of the user interface, and extensive training was undertaken to equip the operators with the skills to use the new system. The installation and commissioning was completed ahead of schedule and below cost, and the explicit benefits were deemed to have been enabled for delivery at start-up. However, when the process was restarted, the yield and cycle time were 15–20% below the historical average. This was a completely unexpected negative benefit of the improved control and the customer was deeply unsatisfied. The process operators also complained that the new system lacked the flexibility of the old one and that was the reason for the poor performance (operations, and not the project, were responsible for delivering batches of product).

Outcome

A review with the operators highlighted that the old system allowed a number of 'work-arounds' to enable the plant to meet production targets that had increased significantly since the original facility went on-line. The new system effectively 'locked in' the process as the technical team understood it and prevented these operational work-arounds. To resolve the issue the Project Team spent approximately three further months revising the automation system software to proceduralize the work-arounds and return the plant to the original performance levels.

Conclusion

The explicit benefits in the project were clear: improved security of supply and improved process control through tracking and tighter control of production variables. The implicit benefits were those expected of any new system – flexibility, capability, speed of performance and maintenance of current performance (yield and cycle time) – generally that a new system would be better than an old one. No attempt had been made by the Project Team to identify and track those implicit benefits. Had they done so, they may have better understood the relationship between project scope and current process performance.

Defining implicit benefits

Implicit benefits realization is about proving that the customer 'contract' has been delivered. This 'contract' aims to make explicit the customers' expectations and agrees the engagement between Project Team and customer. A good 'contract':

- Defines the overall expectations of the relationship.
- Clarifies specific actions or activities which are to be performed.
- Provides the parameters and freedom to act within them.
- Sets agreed goals and objectives.
- Establishes the ground rules for behaviour.

The contract with the customer is usually set up during project delivery planning (Melton, 2008). At this early stage in a project lifecycle the contract aims to establish the expectations that the customer has from the project and it is important to understand:

- *Who the customer is* – this is not necessarily the same as the project initiator. In an earlier case study, the customer was the operations department, but the project was initiated by (and largely engaged with) technical and engineering departments.
- *What their motivation and drivers are* – as the case study in Chapter 4 shows, eliminating the visible steam plume, not saving energy, was that customer's driver.

Tracking implicit benefit delivery

Tracking implicit benefits is not quite the same as tracking explicit ones. By their very nature, implicit benefits are difficult to quantify and therefore difficult to establish metrics for. There are techniques where implicit benefits can be provided with a 'pseudo-value' against which benefits realization can be evaluated.

For example, a sales order business process redesign is intended to provide an explicit benefit of faster order turnaround and an implicit benefit of improving the sales team's product understanding. Using a questionnaire, the sales team's baseline understanding can be scored at project kick-off and the exercise repeated through the project. The Benefit Specification Table (Appendix 12-8) can then be used to track these 'pseudo-explicit' benefits in the same way as any other.

However, this technique is extremely costly to apply to any but the simplest variable. It is also very subjective and there is a risk that the scoring will imply a precision of measurement out of all proportion to reality. It also means that the project may focus on some, but not all, of the implicit benefits which will dictate success for the project. Generally a more holistic way of understanding what will make the customer satisfied is needed to track implicit benefit delivery.

Customer satisfaction

A Kano Analysis (Bicheno, 2004 and Melton, 2008) can be used in its traditional sense to identify and classify customer satisfaction against project scope. The Kano model plots characteristics important to the customer in terms of how fully they are implemented and how satisfied this makes a customer. In a project context, scope and associated implicit benefits can be plotted to understand how delivery will satisfy a customer. The outcome of the analysis is three levels of customer satisfaction.

Must have

If these areas of project scope are not delivered to a basic 'must have' level the customer will have a level of dissatisfaction with the project even if all explicit benefits are delivered. For example, a new way of working in a laboratory removes an important facet of team development because they no longer rotate onto different types of analyses. For the customer the team rotation as a WoW is a 'basic' requirement which he wouldn't expect to have to articulate. Not understanding the WoW caused the Project to generate a 'dissatisfier' that was difficult to remove after project completion.

Without 'must have' scope the basic level of customer expectation will not be realized. For example, a replacement piece of equipment system may be expected to have at least the same basic capability as the old equipment. This describes 'basics' which can become dissatisfiers if they are not met. In Figure 6-8 the customer expects a certain level of communication or he will be dissatisfied – (the stakeholder communication needs to achieve a basic level). However, more communication will not increase satisfaction any further.

More is better

If we have more of this scope then the customer will be more satisfied. For example, if we redesign a part of the process we will improve yield and the customer is satisfied. However, if we can redesign it all he will be more satisfied (Figure 6-8). For these areas of scope it is important to realize when sufficient customer satisfaction has been achieved.

Delighters

If we deliver this scope the customer expectations will be exceeded. For example, a new system may provides novel features which allow the customer to do new things that he previously hadn't thought of but which do deliver value to him. These will not be explicitly requested and may be above the target satisfaction level.

The Kano model is best used at the start of a project to support development of the customer contract and articulation of the implicit benefits. It can also be used to sanity check scope development versus explicit and implicit benefits delivery so that the outcome is both a satisfied customer and a satisfied business (Melton, 2008).

Tool: Customer Satisfaction Analysis Tool

Gauging customer satisfaction is largely about rating perceptions about implied benefits and as such is one of the more difficult aspects of benefits realization. This tool (Table 6-3) aims to provide a structured approach to doing this. It is important to make the effort, particularly in projects with a high

Benefits realization

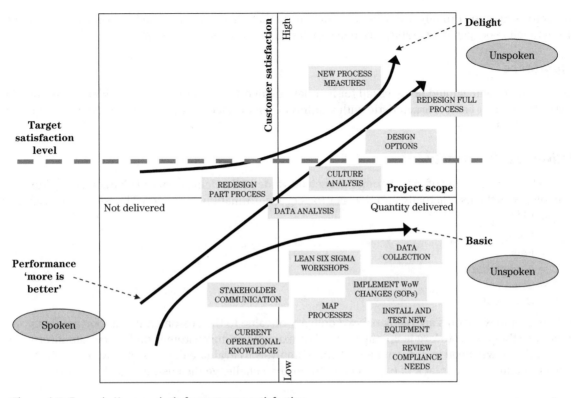

Figure 6-8 Example Kano analysis for customer satisfaction

Table 6-3 The Customer Satisfaction Analysis Tool explained

Benefits Management Toolkit – Customer Satisfaction Analysis Tool			
Project: <insert project name>		**Project Manager:** <insert name>	
Date: <insert date>		**Project Sponsor:** <insert name>	
Customer satisfaction characteristic	**Importance**	**Delivered value/quality**	**Weighted score**
<insert key benefit characteristic>	<insert relative importance assigned by customer>	<insert customers perceived value delivered>	<calculated final score from importance and delivered value>

degree of business change, as dissatisfaction will make the realization of all types of benefits difficult and will ensure that the project continues beyond its expected end point.

Customer satisfaction characteristic

A specific benefit or project area that is of direct interest to the customer. For example improved access to information in a new IT system. There would typically be only a small number of these

Project Benefits Management

characteristics and certainly fewer than 10. Any more than this and either the level of detail is too high or the tangible project deliverables are being included in the satisfaction score.

Importance

Each characteristic should be assessed for its relative importance to the customer. Where possible the score should rank each characteristic with a different importance, starting at 10 for the most important down to 1 for the least.

Delivered value/quality

There should be a rating of the customer perceived value of the delivery of the characteristic. For example, does the user actually find it easier to access information in the new IT system? The rating on this should be:

- 1 = Poor.
- 2 = Fair.
- 3 = Acceptable.
- 4 = Good.
- 5 = Excellent.

Note that these scores are the *perceived* delivery of value to the customer. It is important that you do not use the returns to 'argue up' the rating. For example, if the customer finds access to information only 'fair', you will need to carry out a root cause analysis to evaluate the reason why (lack of training, needs not understood or lack of experience) rather than challenge the customer on his rating.

Weighted score

This metric combines the customer score with the importance score and provides a means to prioritize investigation into areas where the customer is dissatisfied (lowest scores). A uniform low score means you will have serious issues with realizing benefits from the project.

Table 6-4 uses case study data to demonstrate how the tool can be used and interpreted.

Table 6-4 Example Customer Satisfaction Analysis

Benefits Management Toolkit – Customer Satisfaction Analysis Tool

Project:	Automation Upgrade	Project Manager:	Bob Smith
Date:	April	Project Sponsor:	William King

Customer satisfaction characteristic	Importance	Delivered value/quality	Weighted score
System should be as easy to change as the old one	5	2	10
Graphical displays need to be less cluttered and easier to read	3	8	24
Need to be able to easily see why the process is stopped	8	4	32

In this example, the lowest score characteristic (ability to change the system) was the major customer satisfaction issue. In fact, the customer was most satisfied with the visibility of why a process stopped; he just couldn't do anything about it if the recovery logic had not been programmed in. By identifying the specific area of dissatisfaction, the Project Manager was able to do something about it whilst also reviewing its impact on explicit benefits delivery and project completion. In this case additional scope was required to meet customer **and** business needs, in other words the additional scope was a value-add requirement not merely a customer desire.

Benefits sustainability

Sustainability is a way of ensuring that the project is handed over in such a way that the benefits can be realized, all changes remain in place and benefits realization continues. A change which is not sustainable will ultimately not deliver the desired benefits or may cease delivery some time after project completion. For example, a business reporting system was changed to remove printed reports and provide on-line weekly summaries to senior management. Some of these managers were not familiar with on-line systems and so had their secretaries re-create the paper report from the on-line system. This kind of project is a failure as the ultimate customer (user) was dissatisfied with the result and the business rapidly reverted to its previous way of working. The cost reduction (saving in paper and printing time/cost) was not sustained.

As a result of a project there should be some stated vision which describes the sustainable output from the integration of the project, and other associated business changes, into BAU. For example, a project to improve the facilities management processes in a large company headquarters has the vision *'the facility management team will ensure that issues are responded to and resolved efficiently and effectively within agreed time limits'*. The vision should be one that is possible and plausible when considering what the project will deliver (new incident reporting, logging and tracking systems) and what the business will be changing upon receipt of the final deliverable (reduced administrative costs, new SOPs, flexible response teams). The vision should also be worded so that achievement of it can be measured. This supports definition of appropriate sustainability checks and confirms the benefits anticipated from the completed change.

Business as usual (BAU)

BAU refers to the normal, day-to-day operation of the business. It is the aim of the project to change BAU to bring about a benefit. Sustainability means actually integrating the project deliverables into a new normal day-to-day operational model for the business.

An important part of this integration is to ensure that the actual BAU is properly understood and that the project addresses those areas that require change. A Sustainability Plan (Melton, 2008) provides a mechanism to capture all the critical elements of benefit and business change which form the basis of a check on sustainability. The plan should be developed during project delivery planning and may be reviewed as the project delivery progresses (Figure 6-1).

Checking sustainability

A sustainability check should be carried out at (and after) project completion to ensure that the changes are being maintained. There should also be a risk assessment against the vision of a sustainable change:

- What will stop the sustainability checks being positive?
- What will stop the vision of sustainable change being achieved?

- How can high risks be mitigated?
- What is the contingency plan should the risk occur, so that the sustainability check can still be positive?

The Sustainability Checklist (Appendix 12-11) provides a tool to support the identification of sustainability checks and a process of ensuring that change is sustained. It challenges the Project Team to provide a causal relationship between a change that has been delivered and a benefit which is being realized.

Tool: In Place–In Use Analysis Tool

Another way of measuring the sustainability of the change is to look at whether the change is in place and in use. The first (in place) is a measure of the hard delivery of the project (and frequently the point at which measuring stops) and the latter (in use) is an indirect measure of sustainability.

The In Place–In Use Analysis Tool (Table 6-5) provides a means to review and check whether the project deliverable and associated business change is sustainable. It augments the Sustainability Checklist by providing a measure of the business satisfaction of the change. Unlike the Customer Satisfaction Analysis Tool (page 99) which is a tool to measure individual or group satisfaction, this tool is intended to objectively measure whether the business has embedded the change.

Table 6-5 The In Place–In Use Analysis Tool explained

Benefits Management Toolkit – In Place–In Use Analysis Tool					
Project:	<insert project name>		**Project Manager:**	<insert name>	
Date:	<insert date>		**Project Sponsor:**	<insert name>	
Benefit deliverable (in place)	**In place rating**	**Change required (in use)**	**In use rating**	**In place–in use score**	
<insert project deliverable>	<insert rating on status>	<insert business change required>	<insert rating on status>	<calculated final score from in place and in use rating>	

Benefit deliverable (in place)

List the project deliverables that will contribute to the benefit and enable the benefit to be realized. For example, an automated weigh and dispense system for a pharmaceutical manufacturer.

In place rating

Assess whether the deliverable is actually in place and capable of enabling the benefit. The rating levels are:

- 5 = Fully implemented, meets all expectations.
- 3 = Partially implemented, some expectations not met.
- 1 = Incomplete, not capable of delivering benefit.

For example, if the automated dispensary system was delivered without a link to the business system, it would score 3 as some of the expectations (around removing the need to manually enter and record data) would not be met.

Change required (in use)

Identify the business change which is required to realize the benefit of the deliverable. For example, re-writing of SOPs to remove the need for manual weigh records to be kept or a change to electronically readable labels so bar-code readers can be used. These are not within the project scope but must occur if the benefit of having the weigh and dispense system are to be realized.

In use rating

Identify if the business change required is actually in use. The rating levels are:

- 5 = Fully in use across the affected business areas.
- 3 = Partial use, some areas fully engaged – others with limited use.
- 1 = Minimal use or extensive work-arounds to complete tasks.

A rating of 3 would, for example, be typical of a situation where some of the operations personnel in the dispensary area use the new system capabilities fully and others either keep traditional paper records, as a check of the system output, or avoid use of the new system.

In place–in use score

The in place–in use score provides a summary of the status of the various deliverables. A high score is indicative of a sustainable change (it is in use), a low score, the reverse.

Care needs to be taken in reviewing the score if the in place rating is low, but the in use rating is high. This may not mean that the deliverable is not in place, but that the use is forced and the level of customer satisfaction is low. Table 6-6 demonstrates the use of the tool on a project to implement electronic batch records.

Table 6-6 Example of the In Place–In Use Analysis Tool

Benefits Management Toolkit – In Place–In Use Analysis Tool					
Project:	Electronic Batch Record		**Project Manager:**	Bob Smith	
Date:	April		**Project Sponsor:**	William King	
Benefit deliverable (in place)	In place rating		Change required (in use)	In use rating	In place–in use score
Paperless batch record	5		Operations personnel input response on screen	5	25
Exception based manufacturing review	5		Manufacturing supervisor needs to review on screen and sign	3	15
Exception based quality review	3		Quality department needs to review on screen and sign	1	3

In this example, the users in the production area (operations personnel) are making direct and effective use of the electronic record system and the supervisors are making partial use (essentially as a consequence of the fact that a number of supervisors much prefer the paper based system and are uncomfortable with the electronic approach). The quality department, however, are making poor use of the system and further investigation showed that this was as a result of a lack of trust in the electronic results coupled with a repeated request for additional information to support the review process. Therefore the change is not being sustained by the quality team and inevitably this will impact the sustainability of the changes within operations.

Ending the project – sustaining the business

Projects should have a finite life structured around the delivery of a business change and the associated benefit. Sometimes, bringing a project to an end is even more difficult than starting it up and often becomes the point at which the money runs out or the physical item is commissioned/ready for beneficial use. However, this is not necessarily the point at which a sustainable change is implemented. A project actually has three distinct elements to completion:

- Demonstration of benefits delivery.
- Handover.
- Disengagement from the customer relationship.

Handover of a project without demonstration of benefits will ensure failure. Handover without disengaging from the customer may mean that the change is not sustainable.

Short case study: IT system deployment project

Situation

The business required a replacement enterprise resource planning (ERP) system as the existing systems were obsolete and becoming unsupportable. The business case was developed on the basis of a like-for-like replacement of functionality and the project was established as a centrally located Project Team with Deployment Team based on individual manufacturing sites.

Following approval, and to avoid the potential for cost escalation, a decision was made during the detailed design phase of the project to use as much of the ERP system vendor standards as possible. This meant that the replacement ERP system was not really like-for-like with the original, and the way in which some of the functionality was implemented was significantly different.

Outcome

The project delivered the required scope on time. However, there were a number of issues that came to light as the project moved towards completion:

- The user experience was quite different from expectations.
- Considerable hands-on support expertise was required to resolve the usability issues.
- The original system had been heavily customized over time to meet different factory requirements and there were repeated calls for the new system to be 'tweaked' to meet local needs.

As a result of these issues, it became virtually impossible for the Project Manager to disengage from the customer and for the project to be closed out. In fact, the Project Team became the support team (as they were the experts) and the expected benefits of the new ERP system could not be sustained without this level of support.

Finally, the transformation of the Project Team into a support group was formalized and a mechanism to manage the requests for change instituted. There was ultimately no real end to the project in many customers' eyes.

Conclusion

The project as originally conceived failed to deliver a sustainable change because:

- There was no clear understanding in the Project Team of the customer (user) expectations with regard to the system.
- The customer did not fully understand the implications of the change.
- As the system was a replacement for an existing ERP, little effort was expended to look at the implications of the business change.
- The project had no clearly defined handover to local operation/support.

Project completion

Apart from the complete delivery of the defined project scope, a Project Manager needs to ensure that the intended benefits are fully realized sustainably or can be accurately forecast to meet this end point. Such a forecast should make use of the metrics defined in the realization plan and present an agreed milestone of benefits realized at handover point (benefits acceptance). It is not necessary to formally demonstrate that the benefits have been realized to achieve acceptance, usually a clear understanding that the benefits will be achieved is acceptable.

Disengagement by the Project Manager from the relationship needs to be planned for in the development of the customer 'contract'. Disengaging means:

- *Deciding that the 'contract' has been delivered* – agreeing that there is no ongoing need for the relationship as related to the project.
- *Closing out the relationship* – agreeing on what has been delivered, sustainability and the success of the 'contract'.
- *Letting go and saying goodbye!*

Planning for both benefits acceptance and disengagement is a part of planning for benefits realization. Within the realization plan and the customer 'contract', a point at which the project can be deemed to have met its explicit and implicit benefits needs to be defined and agreed upon. For example, often in business change projects training and understanding of new systems becomes the obstacle to handover. There is no way that the implicit (and even some of the explicit) benefits can be realized if the customer is not satisfied with the training or understanding and this should have been made clear during planning and incorporated into the delivery.

Following benefits acceptance and disengagement, it is possible for the project to then go through the final phases of:

Project handover

This is the formal agreement that the project is actually finished. This is only going to be possible if the terms of the customer 'contract' have been met, whether or not a formal contract actually exists.

The customer will want assurance that all tangible elements of project scope have been delivered, that implicit and explicit benefits are enabled (or in the process of being realized) and that the project is effectively integrated into BAU.

Project closure

Closure is not quite the same as completion. Closure is the point that the Project Team is disbanded. An important consideration for sustainability is what support mechanisms are in place to bolster the business change and help ensure sustainability. These may be as simple as trained maintenance teams for a new piece of equipment or as complex as an IT management structure (with new people, policies and procedures) to manage change requests following the installation of a new business system.

Project review

An after action review (AAR) is an important learning step for both the organization and the members of the Project Team. There is also a substantial implicit benefit in this review, the lessons learnt from the project will be of value to the organization for the next project it attempts.

Disengagement

Once the Project Team has disbanded, the AAR has been held and the sustainability reviews (either checklists or In Place – In Use Analysis) have been progressed, the Project Manager can finally and fully disengage from the project and therefore from both the sponsor and the customer.

Evaluating success

As most projects are complex and involve many people, it should not be surprising that evaluating success can be difficult. Projects need to be assessed at three distinct levels:

- *The strategic level* – were the strategic objectives met?
- *The project level* – were the project objectives met?
- *The process level* – how did the project go?

It should be evident that success could be achieved at all three levels, at any combination or at none at all. In Chapter 2 (Figure 2-3), the balance between the strategic goals and project goals was examined, and it was recognized that a project which met all of its internal targets could still be judged a failure if circumstances meant it did not deliver its strategic goals. Similarly, a project which delivered its strategic objectives but failed to meet its project objectives may be considered a success as it contributes positively to the overall performance of the company.

The process by which the project was implemented may also be a factor in assessing overall success as it may have significant effects on both how the organization feels about the project and its attitude to future projects. A successful project which was very difficult to implement may be seen less favourably than one which just failed to meet its targets but ran smoothly and without conflict.

Similarly, different stakeholders are likely to perceive the combination of outcome parameters differently and arrive at different views. So there may be no consistent corporate view of how successful any one project has been.

It may also be recognized that a project which fails to deliver on any particular basis may have been a success had it not been for changes in the business and competitive environments during its delivery.

The uncontrollable external changes may have been so severe that it would have been impossible to meet the original targets.

Whilst it is useful to have quantitative methods of assessing performance, preferably using some form of balanced scorecard, it is likely that a consensus view of the success of each project will emerge over time. This may bear no resemblance to what the participants and stakeholders felt during implementation.

Project success is often subjective and most projects are neither total successes nor total failures. There is a danger that attention will be focused simply on areas where the project failed to meet targets and this may get in the way of learning from the whole experience. It is important to concentrate some effort on identifying what has gone right as this may give insights to aid future success. In any event, project benefits management is an important contributor to the view of project success as it maintains the link between the project and the business, at a strategic, customer and user level.

Handy hints

Finish your project!

There are many projects which achieve 99% completion and then fail to complete the final elements. A lack of completion of any of the project scope will have an impact on benefits realization, assuming that the scope has been developed purely to achieve those benefits.

Have an 'active' project handover

Too often handover of the project is a passive activity: where the Project Manager merely completes the project and walks away. A successful handover needs the project scope to be given and received so that the customer contract can be completed.

Have a plan and use it!

Benefits realization and sustainability plans can be helpful documents for all stakeholders at this 'fuzzy' end of a project. They clarify what needs to be achieved, by whom and when, in order for benefits to be delivered and sustained.

Be interested in benefits realization even if it's not your job!

Some traditional projects may define a very hard handover stage beyond which a Project Manager is not expected to be involved. Even in this scenario a Project Manager should be interested in supporting benefits realization.

Be interested in sustainability

What scope is delivered, how and when, can have an impact on whether a benefit is sustained. A Project Manager needs to understand the link between scope, benefits and sustainability in order to ensure that planning and delivery are appropriate and will enable benefits realization.

Measures determine behaviours

Ensure that the benefit and sustainability measures drive the right behaviours, as well as the right actions in the business.

And finally . . .

- Benefits realization should be the common goal for the sponsor, customer and Project Manager and as such should be integrated into all contracts during the early stages of a project.
- If you don't sustainably realize the project benefits then the project has not been successful and the business investment has been wasted. Organizations have limited resources and cannot afford to waste any investment (funds, people, and assets).
- The ultimate goal for a project is to integrate scope into the business, and by doing so realize benefits for the business.
- If you don't measure benefits realization . . . project success is only a rumour.

7 Short case studies

The case studies detailed in Chapters 7–11 attempt to highlight different scenarios which can occur within Stages One and Four. Some demonstrate good practice and others not so good practice.

The following short case studies, (Chapter 7), are focused on highlighting the reasons why benefits management is needed.

Case study A – the 'personal' project

This case study is based on a project which was being championed by a very senior executive team member. Initially it had all the hallmarks of being a 'personal' project, one that the specific champion wanted to push through the system, but which didn't appear to merit delivery.

Equipment utilization project

Situation

A large pharmaceutical site which contains both manufacturing and R&D operations has been undergoing a major operational cost review through benchmarking key performance indicators with other company sites, as well as with other pharmaceutical companies.

For many years the issue of low equipment utilization on site had been championed by a senior executive team member. It appeared that with the pressures on reducing revenue and capital budgets in the coming year, he would be able to get a project approved to improve utilization.

Whilst operations managers could understand that this was one potential area to look at, they were not convinced that it would deliver the financial savings (either stated as possible by the project executive or as required by the site). Most of them also felt that the potential project was only getting approval because it had the very public backing of the specific executive team member – it had become his 'personal' project.

Potential project

Recognizing that the company had an approvals process for all projects, the executive team member developed a business case which he intended to put forward at the next project approval committee meeting. As he had been pushing for this project for a number of years he found it relatively easy to complete the mandatory paperwork (a one-page business case) and duly submitted the document (Table 7-1).

Table 7-1 Example Business Case Tool (project not delivered)

	Equipment Utilization – Business Case
Business Case developed by:	Confidential	**Date:**	Confidential
Project reference number	Confidential	**Business area**	Operations
Project Manager	Not yet allocated NOTE: it was very unusual within this company to develop a business case without having some resources allocated including a project manager	**Project sponsor**	Confidential NOTE: this was the senior executive team member
Business background		The operations area of the business is constantly requesting funds for new equipment, yet the current equipment is heavily under utilized NOTE: no 'hard' evidence was presented to support this statement, the weakness of the problem statement was a major reason why the project was rejected in its current form
Project description		The project will develop processes and procedures to improve equipment utilization across the whole of the operations area NOTE: the exact scope of the project was not defined in terms of how many processes needed to be developed or how the appropriate level of equipment utilization was to be determined
Delivery analysis		To be defined NOTE: no front end or conceptual definition work had been progressed, so the likely level of resources required to deliver this project (internal and external), the required capital and revenue, and the overall project time-line had not been identified. Also no risk assessment had been conducted or any review of dependency issues related to other 'live' projects or activities within the organization
Business change analysis		This project will support increased collaboration across the various units within the operations area NOTE: the ease or difficulty with which this additional collaboration is possible was not discussed, nor were any other potential operational issues with changing the way equipment was utilized
Value-add analysis		This project will support increasing operational efficiencies (revenue saving) and will reduce financial investment in equipment over the coming years (capital saving) NOTE: there was no detailed cost/benefit analysis due to the lack of project definition. The exact metrics which were to be measured to provide 'value-add' at the end of the project had not been defined either
Impact of NOT doing the project		Without this project the operational inefficiencies seen due to the unacceptably low level of equipment utilization in the operations area will continue NOTE: The senior executive member stressed this area when presenting the potential project to his colleagues on the approval committee
Project approved (Value-add or not?)	YES/~~NO~~ NOTE: although initially approved, the robustness of other business cases highlighted the weakness of this business case	**Name of approver and date**	Confidential NOTE: this was the senior executive team member who leads the Project Approval Committee

At the approvals committee meeting the potential project was reviewed first and duly approved. The rest of the committee were impressed with the passion and persuasive rational put forward. The remainder of the meeting progressed, with many projects being proposed for a limited level of resources available. Soon it became clear that the equipment utilization project business case was not as robust as others being reviewed. The distinct lack of data to back-up the proposal and the lack of understanding of what it truly could deliver became obvious. As a result the project champion, the executive team member, proposed that the 'approved' project be stopped until further front-end work could be completed.

Table 7-1 includes notes in bold italic. These were inserted after the presentation of the business case to the approval committee and highlights why this project, although initially approved, was not actually delivered.

The outcome

A small team of operations managers, working closely with the project champion (who took a sponsor role) reviewed the issues (Figure 7-1) so that a robust 'problem statement' could be defined. This enabled them to redefine the potential project as an operational review and improvement project in a specific area of the site (manufacturing of bulk pharmaceutical chemicals). The preliminary scope definition is outlined in Table 7-2.

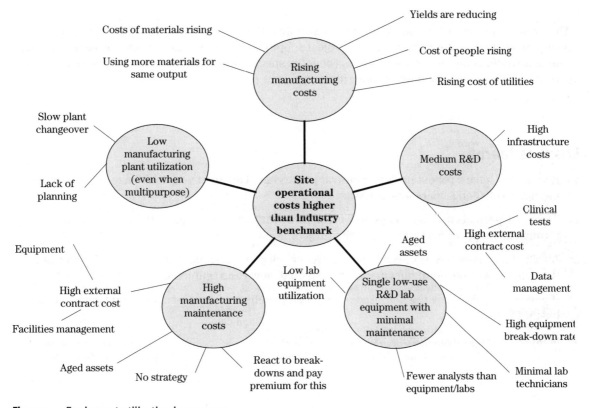

Figure 7-1 Equipment utilization issues map

Table 7-2 Case study CTQ Scope Definition Tool

Benefits Management Toolkit – CTQ Scope Definition Tool

Project:	Operations review and improvement	Project Manager:	Operations Manager
Date:	Confidential	Project Sponsor:	Senior Executive

Benefit criteria	Critical feature	Project scope	Deliverable
Improve output per unit operational cost for plant A	Increased plant utilization	▸ Review all equipment (reliability, appropriate for future products) ▸ Find root causes of low utilization ▸ Use lean six sigma techniques to redesign the facility ▸ Replace aged assets if required	▸ New plant layout ▸ New plant SOP's ▸ Multipurpose plant strategy ▸ New plant assets
	Increased plant change-over speed		
	Reduced plant down-time		
	Reduced materials usage	▸ Track materials use and wastage	▸ Materials tracking SOP

The eventual project which was approved was quite different from the original project. However, it was very explicit in what it would deliver (project scope), how this would be done (project strategy), the value it would generate for the business (benefits specification) and the way the business could sustainably deliver them (business change strategy).

Lessons learnt

- If you don't have a robust project approval process in place then you don't have any objective method to challenge 'personal' projects.
- 'Personal' projects may be completely valid, however sometimes political power may push through an inappropriate or low priority project.
- A robust business case is only robust if it can be appropriately backed-up with real data rather than historical, anecdotal information or personal views and opinions.
- If you don't define how the potential project can benefit your business then you haven't got a business case.
- Make sure that you are solving the 'right' problem.

Case study B – comparing different approaches to business case development

This case study is fictionalized, but is based on an amalgam of situations experienced over a 25 year period.

It compares and contrasts the approaches adopted by three types of manufacturing companies in their attempts to develop multi-purpose/multi-product facilities. It highlights the differences in purpose and understanding of the link between business strategy and the project process (Table 7-3).

Company A

Situation

This company was a successful manufacturer of speciality chemicals. Its customers were generally larger chemical companies who used these products as feedstocks to their processes.

Potential project

The company was concerned that its relatively small production capacity implied a significant overhead on each tonne of production and this, in turn, limited its ability to compete and earn profits. It had long-term objectives to increase turnover and profit margins. However, it found it difficult to justify the expenditure needed to increase capacity. It felt unable to make a proposal to its parent company for investment without having a specific product or set of products on which to base the application.

The local management was also concerned that a generic project scope would not adequately justify either the required investment or the likely returns. They were not confident that such a project could be justified. A final concern was that production contracts were often placed for relatively short periods of 3–5 years, although the plant was likely to have a lifecycle well in excess of this, and would continue to contribute to profits for many years. The management team struggled to find a way of building this long-term value into the project justification.

Outcome

Over the years, the management team formulated several plans for a major expansion of the plant. Most floundered at the feasibility stage and those which were approved had budgets which were incompatible with their stated production objectives. The plant was developed in a piecemeal manner with new features introduced to meet the requirements of specific contracts and the step change in capability was never achieved.

Company B

Situation

This company was the successful operator of a large scale continuous process which used a hazardous chemical. Over time, the company developed a great deal of skill and know-how in handling this chemical. It was also aware that a significant number of high value speciality molecules incorporated this chemical. Whilst it had no experience in batch processing or the equipment types required to produce these products, they had ambitions to build and operate a multi-product speciality chemical facility on their site.

Potential project

Once again the organization found it difficult to develop a business plan to support its request for capital. A major stumbling block here was that the development team was unable to identify an appropriate and robust product slate upon which to base the plan. This made it difficult for the engineering team to produce an outline plant design as a basis for costing.

This impasse continued for several years and it was only after the appointment of a business development director who had experience in a company used to operating multi-product plants, that a viable scheme could be put together.

The business plan was developed on the basis of a list of specific product types, typical feedstock chemical requirements and generic processing equipment. This allowed a contribution based approach to be used to assess the viability of the capital expenditure. This in turn allowed the research, engineering and marketing functions to complete their work on mutually compatible bases.

A consequence of this approach was that the equipment design either needed to be based on a broad but agreed operating basis or left fluid until as late as possible in the project. As the process design was finalized, the range of options on potential products narrowed. For the most part, the selection of the production portfolio was set by the range of conditions and processes the selected design could accommodate. In the relatively few situations where this caused problems, specific decisions could be made on whether to modify the design to meet the required processing conditions. These individual decisions could then be made on the basis of the marginal benefits and costs (taking credit where appropriate for the broadened capability).

Outcome

The project was justified and implemented successfully. The range of processes and product types produced on the plant gradually increased over time.

Company C

Situation

This company was an established speciality chemical manufacturer with a significant proportion of their throughput based on contract production for major companies. They had ambitious growth plans and saw this as a key component of their business strategy.

Potential project

They were aware that in order to meet the growth plans that were built into the overall business plan, they needed to build new plants regularly. However, they also understood that they could not decide on the exact configuration and make up of each plant until production contracts had been secured, a 'chicken and egg' situation.

This problem was overcome by splitting the approval process into two parts. The first part was based on the overall business plan and capital for:

- The building.
- Storage.

- Services.
- Infrastructure (e.g. roads and drains).
- Safety and environmental equipment.
- An agreed set of basic equipment.

It was justified on the anticipated contribution to the company's profit based on a net income per tonne of installed capacity basis. These approvals were made on an annual basis by the main board.

When contracts for production were considered, the costs of modifying the basic plant configuration and any additional processing equipment were identified. This then allowed a separate commercial decision to be made on whether the contract and the implied investment were appropriate. These decisions were made by the site management team within guidelines set by the main board.

Outcome

This staged approach to project approval fitted well with the company's strategy and was understandable to all concerned. It also put approvals at the appropriate positions in the company, strategic level decisions were made at board level and operational ones by the site management team. It also put decisions about new builds on the same basis as the modifications of existing plant needed to suit new or changed contracts.

Table 7-3 summarizes the different approaches and outcomes seen. Having clarity of purpose and an understanding of the business drivers is not necessarily enough to ensure the 'right' action is taken for the organization.

Table 7-3 Comparison of company A, B and C

Comparator	Company A	Company B	Company C
Clarity of purpose	Good	Good	Good
Understanding of business drivers	Average	Average	Good
Willingness to challenge	Poor	Poor then good	Good
Management drive	No drive	Drive from new recruit	Constant drive
Outcome	No action	Delay	Growth

Lessons learnt

- Business plans are like any other plan, they can only be as detailed as the information you have available at the time they are produced. They need to be 'live' and be adjusted as information becomes available.
- The less specific the objectives, the more flexibility is needed in the justification process.
- Some strategic level decisions may need to be made on the basis of belief rather than hard evidence. The business plans to justify these investments will necessarily be less detailed and be based on more assumptions.
- In assessing options to move into new business areas, it is important to weigh the risks of inaction against those of action. The normal project justification process may be inadequate for this level of decision and separate representations may be appropriate.

Case study C – the 'question mark' project

This case study has been used as it provides an example of a 'question mark' project. Had the initial business case been used to assess the benefits, the project would have been deemed a failure, missing many of its more obvious targets (budget, timeline, client confidence). However, the strategic benefits were far broader than the Project Manager was aware. In retrospect, despite the implementation difficulties, the objectives of the organization were met by the project.

Overseas Expansion Project

Situation

The business was an incoming call centre based in English speaking Latin America, with a predominantly US client base. The company employed approximately 1,000 staff. In order to facilitate additional growth and achieve its strategic objectives, the business needed to secure sites outside of its current location. In addition, existing clients have repeatedly enquired about redundancy options in the event of a significant service level interruption, of which the company has none.

The project

The focus of the initial project feasibility work was to align the potential expansion options to the overall needs of the organization (Figure 7-2).

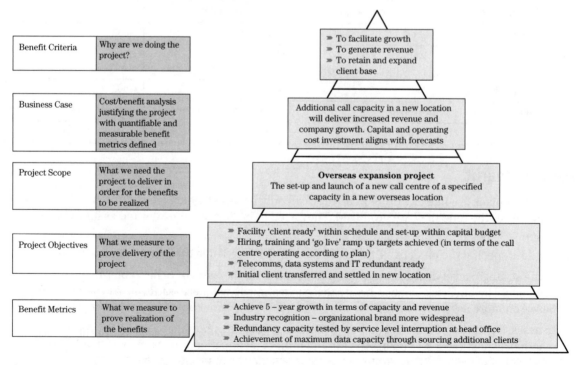

Figure 7-2 Case Study Simple Benefits Hierarchy

During this time a potential site was selected and a potential client portfolio. Both were important factors in the development of the business case which was developed and presented to the board of directors by the Project Manager (Table 7-4).

Table 7-4 Business Case Tool – Overseas Expansion Project

Project Management Toolkit – Business Case Tool			
Project:	Overseas expansion project	**Project Ref:**	XXY123
Business case developed by:	Polly James	**Date:**	End month 1
Project reference number	STL5446	**Business area**	Board of directors
Project Manager	Polly James	**Project sponsor**	CEO – Austin Hibbert
Business background	<ul><li>The business (call centre) has received queries from both existing and potential clients regarding structural, capital and human resource redundancy capability in the event of major service level interruptions</li><li>The business has experienced a sustained period of growth that is expected to continue for at least 5 years</li><li>The FTE (Full Time Employee) labour pool in the current location is saturated</li><li>A strategic goal of the business is to become the 'near-shore' service provider of choice for the North American market</li></ul>		
Project description	<ul><li>To establish an overseas facility, ramping up from 100 to 750 FTEs within 5 years</li><li>To maximize growth/revenue potential in a growing market</li><li>To provide company capacity for structural, capital and human resource redundancy in the event of major service level interruptions</li><li>To drive the business towards the strategic goals of the organization</li><li>To grow the business into a recognized regional brand and become an employer of choice in Latin America</li></ul>		
Delivery analysis	<ul><li>Identify appropriate location (country and site)</li><li>Establish business in host country (legal and financial requirements)</li><li>Refurbish site to meet company and client specifications</li><li>Identify capital requirements including telecommunications, IT, data systems, facility and support services</li><li>Identify and order long lead items such as international fibre optic termination, generator and mission critical data/telcom systems</li><li>Identify, hire and train senior human resources for long-term management</li><li>Identify, hire and train FTE labour pool</li><li>Generate site appropriate training material</li><li>Identify and partner with external service providers such as canteen vendors and transportation services to offer reliable services to FTEs</li><li>Establish FTE payment cycle and acceptable methods</li></ul>**Dependency issues**<ul><li>Project dependent on either expansion of existing client or securing new client. New site must be suitable for requirements of particular client (e.g. location, business type, FTE labour pool)</li><li>Significant initial capital must be accessible within 4 weeks of kick off (approximately U$500,000)</li></ul>		

(Continued)

Table 7-4 (Continued)

	*Project Management Toolkit – Business Case Tool*
Project:	Overseas expansion project	**Project Ref:**	XXY123

Delivery analysis		**Attachments**
		- Milestone schedules developed – broken down into facility (infrastructure, telecoms, IT, data systems and CAPEX assets), and business (hiring, training)
		- Table of CSFs developed (detailing build-out, HR and ramp up CSFs)
		- Risks:
		   ◇ Implementation fails as site/infrastructure/labour not appropriate
		   ◇ Implementation succeeds but cost/revenue (profit per head) ratio poor
		   ◇ Risk of company reputation being damaged if project unsuccessful
		   ◇ Risk of downturn in overall growth and site not needed
		   ◇ Risk of redundancy not being workable

Business change analysis		- This is the most costly project the business has undertaken to date – will create financial constraints in all other areas of the business - Will change WoW as all service departments (IT, HR, accounts, facilities) will have to incorporate overseas site into day-to-day functions - Business will become an international organization and must brand itself accordingly
Value-add analysis		This project aligns directly with the main strategic goal for the organization – increased revenue and therefore increased profit. The capital investment will be paid back within a reasonable time-frame and the operating costs are lower than benchmark for this type of call centre.
Impact of NOT doing the project		- FTE quality may soon impact business as labour pool becomes over saturated in current locations – growth impaired due to labour shortage - Company risks losing new business as there is no service level interruption alternative in place - Company risks losing existing business to larger competitors with service level interruption options - Strategic goals not achieved (to be both the service provider and employer of choice in the region)
Project approved (Value-add or not?)	Yes	**Name of approver and date**	Board of directors End month 1

The business case was approved and the project proceeded. There were many significant problems during implementation, some of which were:

- The client that provided the first 100–200 seats was not the client originally expected. The business case and site selection were completed based on requirements from a different client. The new client had different operating requirements that created difficulties for the site.
- Many assumptions had been made regarding the culture of the country selected for the overseas site. It was thought that there would be more similarities between the original site and the new site than there actually were. This led to significant operational difficulties and a risk of damaging the reputation of the company in the new country.
- The original site selection had been based largely on data gathered from government sources and official promotion agencies. Some of this information proved to be out of date or inaccurate, particularly in areas such as labour, overhead and infrastructure costs. As a result the budget overran significantly in many areas.

- The international telecommunications provider was not able to provide fibre optic links as fast as originally expected, leading to a delay of the first 'go-live' date of one week. This had financial consequences in terms of lost revenue and labour costs, and credibility costs in terms of integrity in the eyes of the client. The original client pulled out of the site after 12 months due to quality and production target achievement issues. However, a second client had been sourced and successfully implemented during the initial year, and a third was imminent. There was a 2 month gap between losing the first client and gaining the third, during which full time employees (FTEs) had to be placed on a retainer. The loss of the client damaged the reputation of the company in the new country. However, it did not damage the reputation of the company in North America, as the client was retained and expanded in the head office location.

Outcome

In the business case the primary benefit was to facilitate growth (more FTEs, increased revenue, retained and expanded client base). In order to demonstrate successful delivery of the project the critical success factors were:

- Delivery of new overseas facility on time and within budget.
- Hiring, training, 'go-live' and ramp up targets achieved (revenue stream in line with forecast).
- Telecomms, data systems and IT redundant ready.

The measures to demonstrate realization of the business benefits were:

- Client and business 'settled' in new location.
- Additional clients sourced and installed on time and within budget.
- Industry recognition.
- Redundancy capacity tested by a service level interruption at head office.
- Achieve 5 year growth and revenue targets.

The project failed to deliver in many of these areas. It was over budget, behind time, unsettled, and at times threatened the reputation of the organization both in the new country and internally. However, both redundancy and industry recognition were achieved.

There was also a significant business benefit that was not included in the original business plan. One of the strategic goals of the business was to become the service provider of choice in the region. The business was privately owned by a small group of businessmen whose ultimate aim was to position the company globally in order to maximize its market worth. The tactic of expanding into the wider region placed the company in a competitive position in the market. Within 6 months of implementation of the project, the board had several interested buyers, and were ultimately able to sell the business for significantly more than had it been a single site service provider.

The Simple Benefits Hierarchy as a tool in the context of the project was not completely appropriate because the ultimate benefits criteria it could support had not been articulated by the business. The project was actually intended to deliver a fundamental shift in the business, rather than address specific capacity constraints. A more appropriate method of summarizing the benefits of this project would have been the use of the benefits scorecard (Chapter 3).

Such a scorecard was ultimately developed based on the complex organizational benefits expected from the project and derived from the overall set of strategic goals Figure 7-3).

Using this scorecard approach it is clear that the emphasis of the project is not on operational issues, but on customer relationship and organizational development. These are strategic business

Project Benefits Management

Figure 7-3 Sample benefits scorecard

goals and much 'softer' than the business case benefit assumptions. Indeed the strategic growth and revenue target 'benefit' is in fact enabled by the project – but cannot in itself be a project deliverable. The 'hidden' benefits hierarchy would have demonstrated that the project was clearly aligned to the strategic goal of 'maximizing the company value' with the ultimate benefit being the sale of the business at that maximum value.

Assessing the softer benefits of customer relationships involves understanding the level of customer satisfaction with the way the project has delivered change to the business. Use of the customer satisfaction tool (Chapter 6) as shown in Table 7-5 would have helped the Project Team formalize and

Table 7-5 Customer Satisfaction Analysis

Benefits Management Toolkit – Customer Satisfaction Analysis			
Project: Overseas Expansion Project		**Project Manager:** Polly James	
Date: End Project		**Project Sponsor:** CEO – Austin Hibbert	
Customer satisfaction characteristic	**Importance** Maximum score = 10	**Delivered value/quality** Maximum score = 5	**Weighted score** Maximum score = 50
Business no longer seen as vulnerable to single failure	6	5	30
Recognized as an industry leader	7	4	28
Business seen as viable in the long-term – capable of growth	8	4	32
Valued by clients	10	5	50

122

understand the implicit benefits from the project and helped guide them through the benefits realization process.

Table 7-5 does not contain the responses from any one person or group, but is a snapshot from customers and key executives in the market place. The scoring shows that in one significant implicit area (valued by clients) the business performs extremely well and in that regard the project has been a huge success!

Lessons learnt

- It can be easier to identify costs and risks than potential business benefits. Many of the people associated with implementation felt that it was an extremely challenging project to execute, and that it failed to deliver what it set out to achieve. Several key resources were lost and the reputation of the business in the host country suffered. Had the primary benefit been more central these issues could have been avoided.
- Don't be afraid to ask questions.
- It could have been possible to avoid many of the implementation difficulties had the primary strategic benefit been identified. As the time line of the primary benefit was not urgent, someone may have been able to 'put on the brakes' at an early stage and analyze whether the new client was a good fit for the new site. Instead, it was assumed that the client request for additional capacity as soon as possible was the over-riding factor, and the project proceeded apace.
- Strategic business goals need a higher level strategic tool to evaluate and understand simple project benefits. A benefits scorecard (linking in the project deliverables to the strategic areas of interest of the company) is a way of identifying the hidden benefits of the project.

8. Case Study One: product storage and distribution facility project

The case study is designed to illustrate the stage gate approach to project sanction and the importance of identifying value in the process.

Situation

Company Z is an established company producing edible oils, related products and derivatives to the UK markets. Its customers are predominantly major industrial concerns who use the company's products in their processes. Company Z has no consumer products and virtually all of its products are distributed by road tankers.

The company has been in existence for over 75 years and has had a number of owners over the last 20 years. It is currently owned by a European based conglomerate with interests in the food, flavourings and pharmaceuticals intermediates market. The group has ambitious growth targets but is cash constrained as a result of its aggressive acquisitions strategy and is currently facing demands for investment from several of its subsidiaries.

Company Z's production site has grown over the years and includes both new state of the art computer controlled facilities and less sophisticated facilities, some dating back more than 50 years. Of particular concern is the product storage and distribution facility. The majority of this is over 40 years old and has grown in an unstructured manner over the years. The unit comprises a large number of relatively small tanks of a range of sizes, standards and specifications. The majority of customers are large industrial concerns who impose their own specifications on the products. This means that Company Z produce a large range of individual grades, many of which are quite similar in nature but need different additives or subtle changes in the manufacturing process. The combination of the number of grades and the range of storage tank sizes means that production planning is very complex, both for the main production plant and for the storage and distribution facility itself. This complexity is compounded by customers' needs to change their orders to suit the demands of their own production processes.

The marketing department is becoming increasingly concerned that the company is losing key contracts as a result of failing customer audits, with the storage and distribution facility being quoted as the prime cause of concern. It is also known that new regulations will be implemented on food hygiene in a couple of years which will accelerate this process. There are concerns about the viability of the site without major investment, but it is felt that it will be difficult to meet the company's investment criteria on this project.

Potential project

Concept review

An initial suggestion to replace the storage facility was tabled at a meeting of the site development steering committee. In the initial discussion it became clear that it would be prohibitively expensive to simply replicate the existing design to modern standards and both the production and planning departments were desperate for a different approach that would simplify their work.

It was also recommended that the project design should build on recent experience at some of the continental sites of the European organization and it was agreed to contact group-level engineers to review their experience and incorporate key learning points from these projects.

Both operations and the instrumentation/control teams recommended that attention should be given to automation and distributed control functions, to ensure that the staffing of the eventual plant was inline with the company's cost objectives.

The quality control representative suggested that the new plant should incorporate both sophisticated online analysis and provide for a high standard laboratory to match the objectives of meeting customer expectations on both quality and visual amenity.

Finally, the human resources representative highlighted the need to comply with recent changes in working hour regulations and meet the company standards on ergonomics.

As the meeting closed, the Engineering Manager was concerned that he had both a hard job in meeting the investment criteria, and meeting the needs of a range of internal parties all requesting features that would increase the cost.

Based on these discussions, the Engineering Manager felt that the overall project would cost in excess of €40 million and suspected that it would be difficult to justify expenditure in excess of €25 million. He also recognized that if the investment could not be justified, the prospects for the site were poor. In discussion with the Site Manager, he agreed to spend €10,000 on a preliminary investigation of the project. He recognized that a key part of this study would be a discussion with functional experts and his colleagues at sites in Europe where improvement projects had recently been completed. His aim was to produce an initial estimate and project programme. He was aware of 'force field analysis' and recognized the need to maximize the benefits, so he decided to contact the Group Marketing Director to get her views (Figure 8-1).

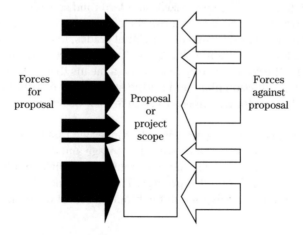

Figure 8-1 Lewin's force field model

Case Study One: product storage and distribution facility project

Initial investigation

As the next meeting of the steering committee was only a month away and the Engineering Manager wanted to submit a proposal to authorize a feasibility review, he needed to move quickly. Discussions with his opposite numbers at the European sites led him to base the initial scheme on the approach used at the site, in Italy, which was most similar to that in the UK. The Italian plant had been installed some 5 years before and at the time had set new standards for the group. This approach also had the advantages that it could be installed simply and effectively on the UK site and required only minor modifications to the production process.

Scaling and updating the costs from this project and making adjustments for the country suggested that the project was likely to cost around €30 million. The timescale for the plant was felt to be around 18 months and the business view was that this would safeguard most of the key accounts.

At this stage, the benefits of the project were considered to be:

- Retention of existing key contracts.
- Increased marketing potential resulting in a higher probability of winning bids in the future.

Cash flows were calculated for both scenarios, investment or no investment, and the internal rate of return (IRR) was evaluated on the differential cash flow. This showed that the project would generate a return of around 7% per annum which fell somewhat short of the investment hurdle of 12% (Figure 8-2).

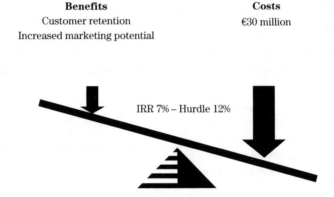

Figure 8-2 Initial review results

Nevertheless, the Engineering Manager felt he had no option but to suggest that a feasibility study should be authorized to see if the gap could be closed. He felt that €75,000 would be appropriate to move the project definition to the next stage.

The site steering committee was concerned that as the expected return was so low there seemed to be an insurmountable gap to be closed, but recognized that the project was crucial to the prospects of the site. So they decided to authorize the next stage and set up a sub-committee to provide additional support.

Feasibility review

A week later, the Engineering Manager attended the group engineering review at the company's headquarters in Munich. He discussed the possible project with his colleagues and was intrigued by a project that had recently been completed in Denmark. This plant processed related products to those on the UK site but used a radically different approach. He recognized that using a similar scheme would put the costs of the project up, but would enable it to deliver some substantial benefits to the main operating plant. The company had an objective of increasing production but the site was struggling to meet even its current targets and there was no consensus on how to achieve the increased targets.

Whilst in Munich, he managed to have a discussion with the Group Marketing Director. She was very supportive of the initiative and suggested that in the event that the UK plant was to close, the group would lose sales of products from other group companies. She agreed to get her team to review the situation and to discuss it with the UK Marketing Director.

Working with a local engineering contractor a revised estimate for the project was completed in 4 weeks and as expected, the costs had escalated, due to both the new engineering solution and the features added by the functional groups. The revised cost of the scheme was put at €40 million. The programme had extended by about 3 months but it was felt that there was scope for reducing this.

Simultaneously, the marketing and operational reviews had reported back, with the result that there were significant additions to the benefits. These were seen to be:

- Customer retention as previously identified.
- Increased marketing potential as previously identified.
- Additional credit for support of group products.
- Operational improvements due to a decrease in the number of product campaigns and an increase in the average duration.
- The ability to produce small quantities of material for trial or bespoke applications without disrupting the main production process.

The result of the feasibility review was that the cost of the project had escalated to €40 million but the IRR increased to 12%, exactly matching the company's investment hurdle (Figure 8-3).

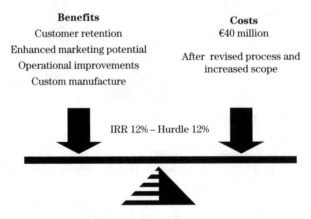

Figure 8-3 Feasibility review results

This information was used to compile a proposal to the board of the UK company to authorize €250,000 to allow preliminary design to be completed. This would form the basis of an application for sanction to the group board. The Engineering Manager was conscious that the project was still looking marginal, but he had had positive feedback from some key people at group level and was confident that the UK board would support his proposal.

Case Study One: product storage and distribution facility project

The proposal was accepted but the board insisted on a tight programme for the preliminary design and requested some additional work to reduce a number of perceived risks. They also reduced the budget for the work by 20% to reflect the shorter time period.

Preliminary design phase

This phase was conducted using a combination of the local engineering team and the engineering contractor who was involved in the earlier phase. Both the Italian and Danish sites were visited and the most appropriate process agreed with the company's technical experts. This was checked for feasibility and reviewed from safety, operations and construction perspectives. This allowed a final scheme to be developed and detailed to the extent that the project risk issues could be resolved and a higher degree of certainty applied to the cost estimates.

The costs moved in both directions. They went up because it was recognized that by adding some additional equipment (to reprocess finished material that failed some quality assurance trials) it was possible to reduce the need to reintroduce this material into the main process. This had significant advantages in reducing variability on the plant which had previously been identified as a major cause of both product variability and throughput. It was also recognized that many of the failed batches could be easily modified to match other product grades. Shortly after this, it was accepted that a modification of this approach could be used to handle the material produced during the transition between production campaigns on the main plant. The combined effect of this was to deliver major contributions towards the key production objectives of throughput, quality and product variability.

The project costs were reduced by the use of a value engineering approach. This had two key elements, firstly, the requirements of functional groups. Production and support groups were critically reviewed to ensure that the proposed solutions were appropriate. Secondly, reviews were held to find alternative, more creative solutions to several key engineering problems. The net effect of this was to reduce the cost without eliminating any essential features of the project.

Having had bad experience of projects where costs were reduced drastically to meet project investment criteria, the Engineering Manager was keen to avoid this approach. Consequently, there was no 'across the board' cost cutting, but only a nominal contingency budget was allowed. It was also decided that if the project were approved, the value management approach would be adopted throughout.

The final budget amounted to some €37 million. The project programme was confirmed and the final revised benefits were stated as:

- Increased marketing potential.
- Additional credit for support of group products.
- Operational improvements due to a decrease in the number of product campaigns and an increase in the average duration.
- Additional production benefits due to improved handling of off-specification and inter-campaign material. Further benefit was also realized with improved reliability and reduced variability of production.
- The ability to produce small quantities of material for trial or bespoke applications, without disrupting the main production process.

On this basis, the project had an IRR of 15% coupled with a robust implementation plan and properly assessed risks. The project was put forward with confidence to the main board who had been sounded out on its acceptability (see Figure 8-4).

Figure 8-4 Sanction application

The project was approved with minor modifications.

Conclusions

The project went ahead and met its functional, cost and programme objectives within close limits. There were relatively few major changes during the implementation phase.

The project was considered a success by all concerned having met both its project and strategic objectives. The production site continues to thrive and has been expanded to manufacture additional products.

Lessons learnt

- It is as important to maximize the benefits of any proposed project as to minimize the costs.
- Bear in mind that your project may contribute to more corporate, site or functional objectives than the ones it is designed to achieve.
- Sometimes adding to the scope will allow additional objectives to be met and increase the attractiveness of the project.
- A value engineering/management approach can be helpful in providing lower cost solutions without removing functionality.
- Beware of adding features which ought to be separate projects justified in their own right, or which logically form part of other projects.
- It is important to prepare the ground for your proposal by lobbying and networking with key people in the organization. They may also give you different perspectives on the project and its justification.

Case Study Two: pharmaceutical facility refurbishment project

This case study has been used to demonstrate the use of the business case template and then to follow the project throughout its life to benefits realization. It highlights the value of a robust benefits specification, both as a scope clarity mechanism and also to challenge whether the project has been successful.

Situation

The case study is based on a project initiated for a major refurbishment of a pharmaceutical manufacturing complex. The facility manufactures a number of (chemical process) stages of a product which was coming under increasing pressure from generic competition, the need for low cost supply for developing countries and fluctuating demand.

The case study is structured around the project lifecycle, but focuses on the benefits activities of each stage. To recap, the four stages of the project lifecycle are:

- *Business case development* – the need for the project and impact on the business.
- *Project delivery planning* – how the project will be delivered.
- *Project delivery* – delivering the project.
- *Benefits delivery* – ensuring that the project delivers the anticipated business benefits.

In terms of the benefits, the four stages essentially focus on the following questions:

- *Identification* – what are the benefits?
- *Planning* – how will they be delivered?
- *Monitoring* – are they being delivered?
- *Reviewing* – is the business seeing the benefit?

Project approval and delivery process

Within the organization, project approval, funding and delivery are very prescriptive processes developed around the typical engineering roadmap:

- *Conceptual design* – initial funding is established by the site and the output of this stage is an estimate and outline benefits case for corporate approval to move to the next stage.
- *Detailed design* – funding (usually a fraction of the total estimate) is sanctioned at divisional level and is used to establish a comprehensive cost and benefits estimate for the next stage.
- *Procurement/construction/commissioning* – full funding is committed at divisional level and the major costs are incurred.

Project Benefits Management

➤ *Customer acceptance/handover* – marks the point at which the project ceases direct involvement and the site assumes operational responsibility. This is also the point at which an after action review is undertaken to evaluate project delivery and confirm the project met its target deliverables and benefits.

Note that although costs are sanctioned at a divisional level, the site is actually charged the depreciation on the capital invested and is therefore very heavily involved in the project financials.

The central theme running through the process is the identification and quantification of benefits at each major stage.

Project selection

The physical assets at the facility (e.g. vessels and pipework) were in good condition, and a preliminary analysis of the production process showed that major contributors to production costs were:

➤ One key raw material.
➤ Labour costs.
➤ Energy for process heating/cooling.
➤ Conversion efficiency for the finished product. This is highly variable, as shown in Figure 9-1:

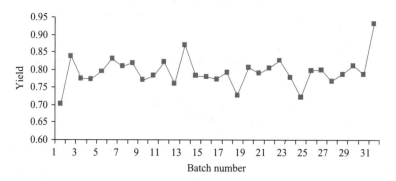

Figure 9-1 Process conversion efficiency chart

Initiatives are in place to address the procurement costs for raw material and energy. A project to reduce labour costs and improve yield and consistency through the use of automation was selected for further investigation.

Conceptual design

A study was undertaken to evaluate the feasibility, benefits and cost of automating the complex. This was a large scale study taking approximately 3 months and involving a team of 6 engineers from various disciplines, as well as local site personnel. The study looked at:

➤ Process operation (through evaluation of, e.g. pipework and instrumentation drawings (P&IDs), standard operating procedures (SOPs)).
➤ People (through interviews across all site personnel).
➤ Problems (e.g. recurrent errors, waste, losses).

A concept design was developed with the following key deliverables:

- *New instrumentation* – the existing instrumentation is dated and inadequate to support automation.
- Process control system providing full automation.
- *Electronic data capture and recording* – to replace current manual methods of recording important information.
- *Electronic Batch Record System (EBRS)* – to replace the existing paper record system.

The mapping of these deliverables against benefit areas is shown in Table 9-1:

Table 9-1 Impact of deliverable on benefit areas

Benefit	Deliverable			
	Instrumentation	**Control**	**Data capture**	**Electronic batch record (EBRS)**
Yield	*High* (better measurements)	*High* (reduced variability)	*Medium* (ability to diagnose after the event)	*Low* (is a post-event record)
Cycle time	*Low*	*High* (removal of human induced delay)	*Medium* (ability to diagnose after the event)	*Low* (is a post-event record)
Direct labour	*Low*	*High* (removal of people)	*Low*	*High* (significant reduction in entry/checking time)
Indirect labour	*High* (reduced maintenance)	*Medium* (removal of some people, but requirement for additional support skills)	*High* (access to external data for investigation)	*High* (significant reduction in entry/checking time)

Benefit estimation

The benefits estimation activity involves expanding the impact assessment (Table 9-1) to explore the detailed benefit in each key area (Table 9-2). The benefits definition should then be developed through formal specification: identifying Specific, Measurable, Appropriate, Realistic and Time-based (SMART) metrics (Table 9-3).

The output of the study was a cost and benefit summary to ±30% (with comprehensive backup detail) outlining the potential for the project. This output was used to justify funding to move to the next, detailed design phase.

Detailed design and project delivery planning

The aim of this phase of the project is to complete the detailed design to the point at which project procurement and implementation can start, and to establish costs to ±10%, allowing full funding to be requested and the project formally initiated.

Table 9-2 Benefits estimation for the project

Benefit type	Benefit estimation
Yield improvements	The process control capability will have a step-change in performance via the use of smart instrumentation and the new control system. This will significantly reduce, if not eliminate, process variations in the following unit operations that have been known to be critical to product quality and yield efficiencies: - Vessel jacket services control - Temperature control of vessel content particularly during reaction, distillation and crystallization - pH adjustment - Solvent metering Product Y Stage 3 – the current yield efficiency is 75% weight/weight (w/w). The best ever yield achieved is 80%, and the variance averages ±4%. It is anticipated that the yield efficiency can be increased by 2% w/w initially with the use of process automation. With the exploitation of the electronic data capture and analysis tools and further fine tuning of the automation system, yield can be further improved by another 1% w/w to give a final yield efficiency of circa 78% w/w.
Manpower reduction	Manufacturing documentation will be captured and maintained electronically. The current manning level supporting the manufacture is 5 operators per shift, with 5 shift teams. A comprehensive evaluation of the operation was undertaken, mapping operations against automation system capability. This exercise concluded that the manning level could be reduced from 5 to 4 operators per shift during the initial phase of implementation. When the system is in routine operation, the number of operators can be further reduced to 3 operators per shift. This represents a saving of 10 operators or a 40% reduction in full time equivalents (FTEs). Experience of other sites implementing such technology have demonstrated approximately 50% reductions in headcount. With the introduction of electronic batch records, process analysis and the asset management system, a further reduction of two FTEs can be expected from the quality department, and the technical and engineering departments.
Cycle time reduction	The processing time of the following rate-limiting process steps can be shortened with the implementation of standard software control schemes, similar to those carried out at other sites: - Batch distillation - pH adjustment operation - Solvent metering operation - Filtration - Drying end point control The time reduction for each of the above unit operations is established with reference to the experience gained from the use of a standard software control scheme in other sites. In addition, due to the implementation of computer sequence control, some waiting time in the production process can be reduced or eliminated. The process mapping and equipment loading methodology was applied to examine the overall impact on cycle time. It concluded that the cycle time of product Y could be reduced from 72 to 65 hours. This represents an increase in production throughput of approximately 10%.

Table 9-3 Project Benefits Specification

Project Management Toolkit – Benefits Specification Table

Project:	Facility Refurbishment		Date:	June	
Potential benefit	**Benefit metric**	**Benefit metric baseline**	**Accountability**	**Benefit metric target**	**Area of activity**
Increase in yield	Average yield over a campaign	75%	Operations	78%	Instrumentation Control
Reduction in cycle time	Average end-to-end processing time	72 hours	Operations	65 hours	Control Data Capture
Reduction in direct labour	Full-time equivalents	25 (5 shifts at 5 persons per shift)	Operations	15 (5 shifts at 3 persons per shift)	Instrumentation Control Data capture EBRS
Reduction in indirect labour	Full-time equivalents	5 (2 shifts at 2 persons per shift, plus 1 general technician)	Technical engineering	(2 shifts at 1 person per shift, plus 1)	Instrumentation Control Data capture EBRS

The detailed design phase essentially took the conceptual design and expanded the specifics to establish costs from customary norms. The costs were therefore established relatively quickly. However, elaborating and justifying the benefits was significantly more difficult.

The implementation of this type of technology at the site was new (the use of electronic batch records was novel within the division), so the site personnel review of the benefits was very extended (almost 6 months), involving repeated value engineering and challenges.

From the benefits map it is clear that the instrumentation and control components of the project appear to deliver significantly more hard benefit than the data capture or EBRS components. The novelty of the technology (particularly EBRS) also introduces an element of risk that the project costs could escalate or that the benefits would not be achieved.

Within the project, a formal value engineering exercise was then undertaken to challenge the extent of the instrumentation required to deliver the benefits (replace only the poorest performing instrumentation) and to challenge the viability of the data capture component and EBRS.

The challenge with the data capture and EBRS components was much more of a 'soft' issue than a 'hard' project deliverable. There was significant resistance to the change in ways of working (WoW) and the potential regulatory impact. To address these, a series of workshops were carried out during the detailed design phase. The workshops covered the following:

- *Technology awareness and understanding* – a fundamental pre-requisite to looking at WoW.
- *Business process* – WoW and the impact on personnel.
- *Regulatory* – what will the impact be from an internal and external perspective?

The workshops took place both with local site personnel and, especially in the regulatory area, with divisional representation.

The principle aim of the workshops was to establish that the technology worked, that the change to business process was achievable and that there were no obstacles to delivering the benefits.

Ultimately, the data capture and EBRS elements were considered essential in light of some of the anticipated soft benefits. These included:

- Improving the potential of the facility for new product introduction.
- Enhanced compliance and traceability.
- Support for process analysis and improvement.
- Enabling remote support of the facility.

Whilst not delivering direct benefit to the project, these soft factors did in fact meet previously unvoiced stakeholder needs, through the positioning of the facility for additional business within the company.

This phase of the project highlights two important aspects during benefits planning:

- A clear understanding and agreement from the key project stakeholders in what and how benefits will be delivered is essential. Otherwise there is a very high risk that during project delivery one or more of the benefit areas will be sacrificed to short-term project expediency.
- There is a need to look beyond the direct benefit areas to try and establish other 'soft' or 'hard' benefits that could drive stakeholder acceptance and support.

At the end of the detailed design phase, a comprehensive business plan was completed and submitted for formal approval.

Case Study Two: pharmaceutical facility refurbishment project

Business case

Note that the title page and table of contents has been deleted and only the main text of the completed business case appears in Table 9-4. The template introduced in Chapter 5 has been used as the basis for this document (Appendix 12-5).

Table 9-4 Facility refurbishment business case

Business Case Template – Executive Summary
1 Executive summary Product Y is the active ingredient in a number of anti-infective compounds. It has been manufactured in Area 14 (A14) for a number of years and is the single source of supply for this active ingredient. The facility is a manually operated plant with a mixture of pneumatic and electric controls, and production staff are responsible for ensuring smooth operation of the plant and control of the process. Consequently A14 is a labour intensive facility and process control parameters are susceptible to variability. In support of the corporate challenge of delivering product cost of no greater than £50/kg of product Y, and in combination with other corporate initiatives, this automation proposal was identified to reduce product Y cost. The proposal will deliver benefits in terms of increased yield, reduction in manpower, decreased cycle time and lower utility cost. The combined benefits amount to £1.1 million/annum at the anticipated demand and will contribute a 2% reduction to the cost of making product Y next year. Payback is 1.6 years with an internal rate of return (IRR) of 32%. In addition to the financial benefits, the proposal will deliver a technology platform that can be expanded to deliver future savings at reduced investment. It will also revitalize an important but aged manufacturing asset. To date £250 000 has been authorized to undertake design and engineering suitable to create a ±10% cost estimate. Total project cost is estimated as £1.76 million, full funding authorization of a further £1.51 million is requested to facilitate implementation with the project scheduled for completion in 12 months.

Business Case Template – Introduction

2 Introduction

The existing A14 plant is essentially manually operated with a mixture of pneumatic and electric controls.

The project will introduce a step change in the way we operate plant control systems and increase the visibility of process parameters, through upgrading existing instrumentation and installation of new, SMART (self-diagnostic) instrumentation. Simultaneously a Distributed Control System (DCS) and Manufacturing Execution System (MES) will be installed and implemented.

A14 is divided into two production areas configured as follows:

- *East:* product Y stage 3.
- *West:* product Y stage 4.

The automation solution will be implemented in A14 East initially and focus on the process steps for product Y stage 3. The principle components are a DCS layer and MES layer.

The DCS layer provides a means to automatically control the batch processing on plant through a combination of sequential and regulatory controls.

The MES layer provides a level of intelligence to the DCS enabling process visibility, reducing variability and improving overall manufacturing performance.

Business Case Template – Project Scope and Organization

3 Project scope and organization

3.1 Vision

The objectives of this project are to reduce product cost by implementing modern automation in A14, and to confirm improved performance and benefit delivery of the technology.

The automation model can then be applied to other processes and streams in A14 for realization of additional benefits.

3.2 Implementation strategy

The project will be implemented in a phased manner as follows:

- *Instrumentation replacement* – to be carried out during routing plant changeover.
- *Field wiring installation* – parallel activity.
- *Control system installation* – parallel activity.
- *Hook-up and commissioning* – during planned summer shutdown.

This approach will minimize the impact on operations and provide an opportunity to build the infrastructure. In the event of delays to the commissioning activities or failure of the control system, the capability to fall back to manual operations will exist.

3.3 Deliverables

Hardware

- Installation of SMART instrumentation and upgrading the existing instrumentation on all vessels, filters and dryers deployed for the product Y process to improve the following unit operations:
  - Temperature control for jacket services.
  - Temperature control for vessel contents.
  - Control for rate of batch distillation.
  - Control for phase separation and pH adjustment.
  - Control for multi-solvent metering.
  - Control for filtration and drying operations.
- Installation of nitrogen blanketing systems for all vessels in the East area to enhance safety protection; and modification of pipe work for the installation of new instruments.
- Installation of a DCS equipped with field based input/output technology as well as the control room and rack room.
- Installation of IT servers and network for the MES.

Software

- DCS software.
  - Generic control modules and equipment modules for multi-product applications.
  - Sequence control software for product Y stages 3 and 4 as well as plant cleaning.

<Project Title>: Project Business Case – Version xx

- MES software.
  - Production information for data capture and storage.
  - Process improvement tools.
  - EBRS.

Excluded from project scope

To avoid substantial rework from the later implementation of the corporate business system, a decision was made not to develop an interface.

Existing assets

This proposal will not result in the write off or disposal of existing assets.

3.4 Stakeholders

Stakeholder	Effect
Engineering	High reliability of equipment More advanced diagnostics
Production	Improved process control reducing manning requirements
QA	Robust process control. Reduced human intervention resulting in enhanced compliance
Technical support	Improved process diagnostic capability

3.5 Related projects

There are no specific projects related to this one. Further cost reduction initiatives are underway through an evaluation of the procurement of material and engineering of the process.

3.6 Organizational impact

There will be a fundamental change in the way the A14 plant will be operated in the future. This will require the following to be considered:

- Training and skills of operators.
- Training of maintenance and support personnel.
- Training of technical support personnel.
- Revision of SOPs, including:
  - System operation.
  - Batch record review and approval.
  - Change control.
  - Calibration.

A Business Change Manager will be nominated from the operations department to co-ordinate the training and SOP activities required.

3.7 Resources

Critical resources required for this project include extensive involvement by operations and technical support staff to ensure specification accuracy and implementation efficiency.

In addition to the internal resources, key principal suppliers from the DCS and MES vendors will be required to supplement the team on a full time basis. The installation, testing and commissioning plan will be prepared and executed by an appointed commissioning manager to ensure that this critical activity meets the production downtime window available.

3.8 Project management framework

The A14 site project engineering group supported by corporate engineering will manage the project.

3.9 Key issues

Issue	Why the issue is important	Plan to deal with the issue
Training and experience of operations staff	System will require new operational expertise	Establish training system with control system vendor and develop training plan

3.10 Critical assumptions and constraints

This proposal assumes sole access to the plant during the summer shutdown (3 weeks) for the purposes of changeover and commissioning of the system. The plant shutdown window is a fixed constraint on the timing of the project.

Benefits are calculated on the basis of forecasted production demand.

Business Case Template – Benefit and Cost

4 Benefit and cost

4.1 Benefits

Financial premise

- Demand profile based on a long-term volumes forecast provided by the Manufacturing Director.
- Yield and cycle time improvements are based on last years budget yield and cycle time.
- Benefits calculated on projected yield efficiencies, manning levels, and cycle-time are shown in Table 9-3.
- The overall project cost benefit analyses are summarized as follows:
  - The IRR is 32%.
  - Payback of this investment is 1.6 years after project completion.
  - IRR will increase significantly if the asset utilization window generated is used for manufacture of new products or increased volume of existing products.
- The impact on product Y unit cost is:
  - The overall cost impact of the improvements on unit cost of product Y is a 5.3% reduction of the cost of making product Y next year.
  - Benefits amounting to approximately £1.1 million at forecast rates will contribute a 2% reduction to the site's total cost of goods next year.

Additional benefits – non-tangible

In addition to the measurable benefits highlighted above, there are many associated non-tangible benefits that can be derived from automation, these include:

- Enhance A14 as new product introduction site.
  - Enable seamless technology transfer.
  - Process information and analysis tools enable efficient process study, investigation and trouble-shooting, thus releasing technical resources.
- Secure and compliant supply.
  - The safety standard of the facility is upgraded with the installation of the nitrogen blanketing system and various safety interlocks and diagnostic tools.
  - Increase reliability of instrumentation.
  - Enhance good manufacturing practice (GMP) compliance via the removal of human variability in the process control.

4.2 Costs

Project cost estimate

The total project cost is estimated to be £1.76 million, with ±10% accuracy. The breakdown of cost is outlined below:

Direct costs	Cost (£)
Equipment	350 184
Installation subcontracts	200 893
Automation	478 348
External services	
Engineering and procurement	475 486
Third party commissioning and vendors	45 871
Internal services	
Staff	108 330
Consultants	21 814
Commissioning	24 178
Other costs	54 895
Total project cost excluding contingency	**1 760 000**
Contingency	176 000
Total estimate including contingency	**1 936 000**

Project direct costs account for 52% of the total project cost, indirect are 38%, with a 10% contingency accounting for the remainder. There are no major items of process equipment or materials to be purchased and project scope is predominantly the automation of an existing manual facility. In this respect, the ratio of direct to indirect costs reflects the high component of specialized software design effort required.

- Contingency, growth and escalation – A 10% contingency has been included in the project cost estimate.
- Related revenue expenses – Annual revenue expenses for maintenance contracts and software support are estimated at 10% of the procurement costs for the automation equipment – £48 000.
- Operational costs – There will be a reduction in operational costs as a result of this project.

Business Case Template – Schedule, Risk, Alternatives and Appendices

5 Schedule

Phase	Forecast schedule (estimated)
Feasibility study	3 months from concept
Scope development	3 months from concept
Project approval	3 months from scope
Completion of detailed design	6 months from approval
Automation design and testing	12 months from approval
Implementation for product Y stage 3	15 months from approval
Implementation for product Y stage 4	18 months from approval
Project completion	20 months from approval

6 Risk

- The project must be implemented within tight shutdown windows with minimum disruption to plant processes/security of supply. To mitigate against this risk, extensive installation and pre-testing will be employed before shutdown.
- The project delivery strategy is fully integrated with the product Y supply plan. Implementation will be carried out in two parts – first with product Y stage 3 plant in East area followed by product Y stage 4 plant upon successful proving of stage 3 implementation. Product Y stage 3 will be manufactured in the West area while the project is implemented in the stage 3 plant in the East area.
- The concept of full automated control is new to A14. The project would therefore be implemented in phases to provide a step by step introduction of new concepts and WoW for the operators.

7 Alternatives

7.1 Identification of options

The following options have been considered:

(a) Do nothing – retain the plant as it is.
(b) Carry out minor control system implementation in specific areas (reactor and distillation control).
(c) Full scale implementation of control system.
(d) Full scale implementation of control system with electronic data capture capability.

7.2 Comparison of options

Option	Comment
(a)	Benefits associated with yield, cycle time improvement, manpower reduction, utilities or new product capacity would not be realized. Plant would remain under-utilized.
(b)	Some benefits of improved operation of critical process steps would occur – approximately 20–30% of the identified benefits in yield and cycle time. Training would still be required for operators. Additional maintenance support would be required. No other benefits would accrue.
(c)	Benefits of improved operation would be received (direct labour, yield and cycle time). Ability to conduct process analysis and improvement activities would be little better than current. Less favourable for new product introduction site. Ability to move to electronic batch records is not possible.
(d)	All identified benefits achieved.

7.3 Recommended option

Option (d) is recommended. It is the only one that provides the performance benefits and enhances the capability of the facility for future new product introductions, an important intangible benefit for the site.

8 Glossary and appendices

- *P&IDs*: Piping and Instrumentation Diagrams.
- *SOP*: Standard Operating Procedures.
- *EBRS*: Electronic Batch Record System.
- *DCS*: Distributed Control System.
- *MES*: Manufacturing Execution System.

Delivery

The project was ultimately sanctioned as envisaged at the conceptual and detailed design phases and the project moved into the delivery phase.

One of the most important parts of the project delivery planning was the insistence that a cross-divisional team be put in place and that core elements of the solution were designed for re-use across the business. This had a significant effect on mitigating against the perceived technology risks, but it did increase the project complexity by adding external demands onto the local project delivery.

The project itself was split into three linked elements with three separate suppliers:

- *Instrumentation replacement* – local instrument technicians.
- *DCS* – systems integrator using hardware and software purchased from a DCS vendor.
- *MES* – project software and implementation directly from the MES vendor.

Co-ordination and management of the whole Project Team was at the site level. Divisional expertise was drafted into the project in a consultancy role – though without specific project authority. One of the key objectives of this consultancy role was to answer technical or business change questions and to ensure that the cross-divisional re-use intent was met.

The informal nature of the consultancy role is one which could have created major problems for the project. Typically an external group without a formal role on the project would be severely challenged. In this case however, the external members had a well established relationship with both the site Project Team and the specialist suppliers being used in the project. They therefore acted as an expert group for issue resolution when questions about intent and approach were raised, and were able to supply insight into the overall aim of the project without becoming lost in the detail of delivery.

Segmenting the project delivery into the three elements raised a number of challenges for the team. Although the instrumentation and control system pieces were linked, there remained the issue of the MES portion being potentially detached from the project and deferred. This problem manifested itself early in the project, when there was a need to refine the technical design of the interface software between batch record system and control system. If this component was not completed as anticipated, then the whole MES portion of the project would have been in doubt.

This issue illustrates the benefit of having external, divisional support in the Project Team. The issue was taken outside the project and addressed directly with the software supplier in collaboration with the control system vendor (who was not directly involved in this project). This resulted in a rapid resolution of the technical issue and the creation of a robust, reusable interface to meet the project testing timescales. Had the issue remained within the local project, it is possible that there would have been no easy resolution and the issue could have derailed the attempt to use electronic batch records by creating mistrust of the technology.

The key benefit of segmenting the project was that the implementation allowed the site to change over in a phased manner, rather than via a 'big-bang'. In the context of the industry, phasing a change and maintaining regulatory security is of great importance.

The first project phase completed was the instrumentation and DCS. This was commissioned on one process stage before any other part of the project was completed and allowed the site team to gain confidence in the system.

The overall project was completed on time and within budget.

Benefits realization

The benefits specification identifies expected hard benefits in:

- Increase in yield.
- Reduction in cycle time.
- Reduction in direct labour.
- Reduction in indirect labour.

The phasing in of the technology, plus the extensive workshops and divisional support provided meant that the hard benefit targets were either achieved or exceeded, as shown in the Benefits Tracking Tool outlined in Table 9-5:

Table 9-5 Benefits Tracking Tool

Project Management Toolkit – Benefits Tracking Tool

Project:		Facility Refurbishment	Date:		Project End
Benefit metric		Baseline	DCS commissioning	MES commissioning	Target
Average (%) yield over a campaign	Plan	75%	77%	78%	78%
	Actual	75%	78%	79%	79%
Average end-to-end processing time	Plan	72	68	67	65
	Actual	72	70	68	66
Reduction in direct labour	Plan	25	20	18	15
	Actual	25	22	18	15
Reduction in indirect labour	Plan	5	5	4	3
	Actual	5	5	4	3

In addition, some significant unanticipated benefits were achieved. Two key ones were:

- Increase in plant capability (sigma level) from approximately 2 to 5.
- Dramatic reduction in batch record review and approval of approximately 75%.

Intuitively, these benefits should have been expected as a consequence of the project. The replacement control system eliminates variance in the process and thereby increases capability. The electronic batch record, by being 'right first time' reduces the need (and time required) for rework of documentation. However, the extreme nervousness about the technology and its impact on the business meant that these aspects were not pursued in the original analysis – the justification was considered unreasonable and as a result the benefits case was too conservative.

A second issue over benefits realization is associated with the recognition of 'cause and effect'. Changes in process performance (yield and cycle time) and labour usage can often be attributed to causes other than the project's direct deliverables. For example, changing the business process reduced the effort required to run the plant and thereby reduced the manning level, or alternatively, reviewing the chemical process forced the correction of long standing issues and improved cycle time. In both these examples it is easy to credit the benefit to another area – especially for those people who are not sold on the project, the technology or the change.

An appropriate way to link the cause and effect of project and benefit, whilst accounting for the inevitably 'fuzzy' areas is to expand the benefits map to include the critical success factors (CSFs) of the project (and make sure that those CSFs reflect the wide range of changes necessary to achieve the benefits). Figure 9-2 illustrates how some of the benefits are linked:

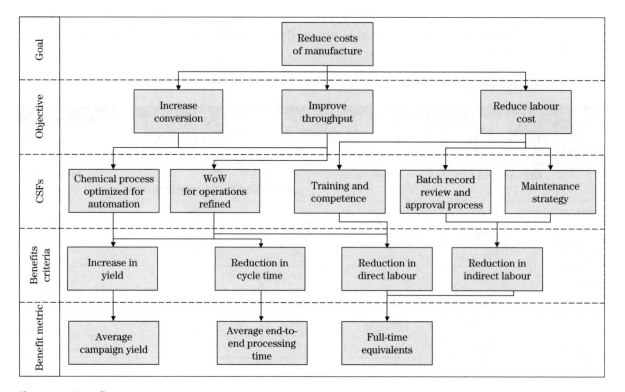

Figure 9-2 Benefits map

Case Study Two: pharmaceutical facility refurbishment project

Conclusions

There are three major conclusions arising as a result of this project:

- *The review, evaluation and challenge exercise was far too long* – in effect 6 months of project benefit was forgone. This was largely due to the reluctance of the site team to accept the expert view of the benefits. Better training and more examples of actual performance in practice would have helped here.
- *A number of benefits accrued that were not even considered at the concept stage* – these included a significant improvement in process capability (in the six sigma sense) and a dramatic reduction in particle size variance in the finished product. Concerns about the technology contributed to this – the novelty of the project was more limiting than expected.
- *Attribution of benefits to the project became difficult* – it is important to identify all of the issues that contribute to the success and link to the benefits as early as possible.

Lessons learnt

- Ensure that all the benefit areas are included in the benefit evaluation.
- Understand the other business drivers (e.g. making the facility a desirable site for new business).

10 Case Study Three: organizational change programme

This case study has been used to demonstrate the value of using the Benefits Management Toolkit within a project to set up a benefits management process within an organization. This project also highlights the value of early benefits concept development and then the completion of the process via robust benefits tracking during realization.

Situation

A large facilities management organization was finding it increasingly difficult to compete in what was becoming an over crowded and highly competitive market. They reviewed their business strategy and concluded that they had four key strategic goals to achieve if they were to regain their former place as market leaders. They needed to be viewed externally as:

- Cost competitive.
- Internally efficient.
- Externally customer focused.
- Providers of quality facilities management services.

The business management team concluded that they would have to make major changes to the operation of the business, although any change would have to align with one or more of the four strategic goals. This level of business change was new to the management team and they quickly recognized that the first project would have to be the set up of an appropriate system to manage all the change.

Business case

The potential project under review was that the business management team needed a process to adequately sift through change ideas and then quickly gain an understanding of how a portfolio of projects based on aligned ideas could move them towards their strategic goal. They appointed a Project Manager who worked with them to produce the project path of critical success factors (CSFs) (Figure 10-1) and also a Benefits Hierarchy (Figure 10-2) to show direct alignment between the proposed project scope and the strategic goals.

It was clear from the discussion with the management team that there was an expectation that one of the benefits from this project would be a measures system that adequately indicated to them their progress towards the strategic goals. However, exactly how this was to be achieved had yet to be clearly defined.

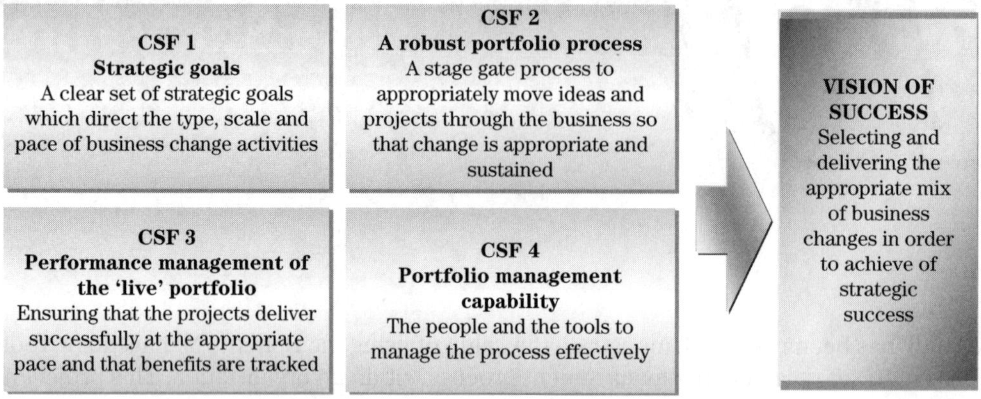

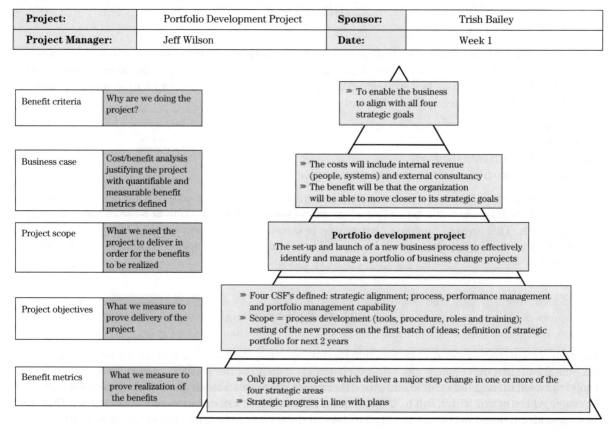

Figure 10-1 Portfolio development project path of CSFs

Project:	Portfolio Development Project	**Sponsor:**	Trish Bailey
Project Manager:	Jeff Wilson	**Date:**	Week 1

Figure 10-2 Portfolio development project benefits hierarchy

To support further scope definition before formally starting the project, the Project Manager held a workshop with the management team in order to better understand their expectations and how this would deliver the required benefits. The initial Kano Analysis highlighted the critical features they expected to see as deliverables from the project (Figure 10-3).

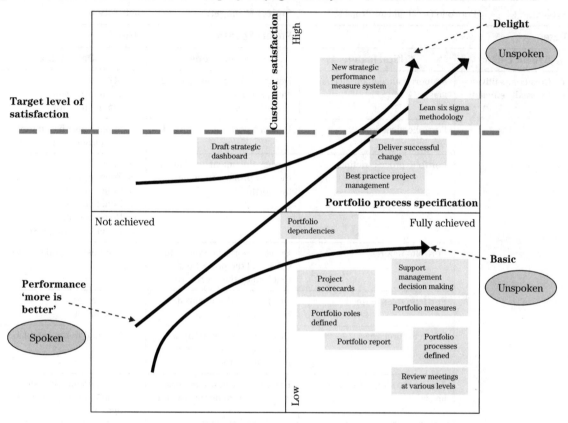

Figure 10-3 Portfolio development project Kano Analysis

The session encouraged the management team to talk about the critical features of the new portfolio process and in doing so they were able to articulate:

- Basic features that they were expecting but had not previously discussed in any detail.
- Features previously discussed which could be delivered to a number of different levels. Clearly 'more was better' for these features, but the workshop identified the level at which they were satisfied they would meet business needs (and therefore their needs).
- Features which they had no expectation of but which would 'delight' them.

In clearly identifying the critical features, the Project Team was able to develop the scope through use of a CTQ Definition Tool (Table 10-1), where CTQ stands for 'critical to quality'. In other words the CTQ features are those which are required to meet the business need. At this stage a benefits map had not been generated so the benefit criteria were assumed to be the strategic goals in terms of ensuring broad alignment.

All critical features were challenged and only two were highlighted as above business needs at this time (those above the target line in Figure 10-3).

Table 10-1 Portfolio development project CTQ Scope Definition Tool

Benefits Management Toolkit – CTQ Scope Definition Tool			
Project: Portfolio Development Project		**Project Manager:** Jeff Wilson	
Date: Week 3		**Project Sponsor:** Trish Bailey	
Benefit criteria	**Critical feature**	**Project scope**	**Deliverable**
• Cost competitive • Internally efficient	Support management decision making	Design and implement a stage gate process	• SOP • Training session • Communication
	Portfolio measures	Design a portfolio scorecard	• Spreadsheet based tool with one page report output
	Portfolio process defined	Design, document and launch the new process to manage the portfolio	• SOP • Training session • Communication • Benefits map • Benefits matrix • Benefits scoring system
	Portfolio report	Design a portfolio system to collate all the project data required by the scorecard	• Spreadsheet or data base
	Portfolio roles defined	Design the roles required based on the process	• RACI chart • Role profile
	Project scorecards	Design the system of project reporting as needed by the portfolio scorecard	• SOP and report template
	Review meetings at various levels	Design the review meetings required based on the process	• Portfolio meeting schedule (within business planning cycle)
	Portfolio dependencies	Define best tool to use	• Portfolio dependency flow chart tool
	Best practice project management	Research into current best practices	• None – the interim Deliverable (research) will be integrated into all other deliverables
• Appropriate quality • Customer focused	Deliver successful change	Test the portfolio process on 'live' ideas	• A populated benefits matrix • Management decisions • A populated scorecard
• All four strategic areas	Draft strategic dashboard	Measures mapping to evaluate what other information the management team need to assess strategic progress	• Draft strategic dashboard tested with data from initial portfolio

Based on the more detailed scope definition a formal business case was developed and internal and external costs approved. Fundamental to this approval was a clear articulation of the project benefits which would be realized and an understanding of the pace and volume of benefit (Table 10-2 and Table 10-3).

Table 10-2 Portfolio development project Benefits Scoring Tool

Benefits Management Toolkit – Benefits Scoring Tool				
Project: Portfolio Development Project		**Project Manager:**		Jeff Wilson
Date: Week 3		**Project Sponsor:**		Trish Bailey
Non-financial benefits score				
Benefit criteria	Maximum score	Benefit level	Benefit impact (%)	Benefit score
Cost competitive	30	1	100	1
Internally efficient	30	1	100	1
Appropriate quality	20	0.5	100	0.5
Customer focused	20	0.5	100	0.5
Financial benefits score			**Total benefit score**	
Benefit type	Benefit level	Benefit multiplier	Score	Comment
Enabler for all types	None	1.0	3.0	Mainly an enabling project

In some regards the Project Manager was developing this business case by modelling the preliminary ideas for the portfolio process which was at this time in 'idea' phase. For instance, project management practices highlighted that benefits scoring (Table 10-2) and benefits specification (Table 10-3) were best practice processes which should be incorporated into the eventual solution.

The benefits scoring system had not been completely defined, so only a very broad score could be established based on alignment with the four strategic goals. However, even this preliminary tool highlighted that business process change projects such as these have organization-wide impact and are more about enabling progress than making it. The actual testing of the process is within the project scope and it is entirely possible that within the test period a change project could be delivered which has major financial and non-financial benefits. However, this is not fundamental to the business case approval.

Based on the work completed thus far the Project Manager developed a draft Benefits Specification Table (Table 10-3) to clarify what measurable improvement this process would have on the business. In order to clarify the link between benefits and scope, the scope in the CTQ Definition Tool (Table 10-1) was segmented into three main types:

- *Area 1 – Design* – all elements of process, tool and role design and documentation as well as the design of the new measures processes.
- *Area 2 – Process Implementation (launch)* – all elements of launching a new business process such as communication; training and safe testing (simulated rather than live).

Project Benefits Management

- *Area 3 – Measures* – portfolio, project, benefit and strategic progress measures.
- *Area 4 – Testing (live use)* – the live use of the process, roles, tools, measures to select projects, build the portfolio and track its performance for a specified period before the system becomes a part of business as usual (BAU).

Table 10-3 Portfolio development project Benefits Specification Table

Project Management Toolkit – Benefits Specification Table

Project:	Portfolio Development Project		**Project Manager:**	Jeff Wilson	
Date:	Week 3		**Project Sponsor:**	Trish Bailey	
Potential benefit	**Benefit metric**	**Benefit metric baseline**	**Accountability**	**Benefit metric target**	**Area of activity**
Internally efficient	Projects approved based on strategic alignment	<50%	Trish Bailey	>90% at launch trending to 100% within 1 year	Area 1 and 2
	Efficient use of resources on change projects	<65%	Trish Bailey	>80% at launch trending to 90% within 1 year	Area 1 and 2
	Balance of change projects per strategic area	None	Trish Bailey	25% in each area based on benefit score	Area 1 and 2
Cost competitive	Benefit score average per project	None	Trish Bailey	>15 unless enabling or strategic	Area 3 and 4
	Financial benefits delivered	<$10,000 per project	Trish Bailey	>$50,000 for each of the four parts of the portfolio	Area 3 and 4
Appropriate quality	Projects which identify service quality changes	None	Another management team member	At least 1 'live' project in the portfolio	Area 4
Customer focused	Projects which improve customer satisfaction	None	Another management team member	At least 1 'live' project in the portfolio	Area 4
	Strategic progress	None	CEO	15% in year 1, 25% in year 2	Area 3

Delivery

Following formal project approval the Project Manager formed a small team with some external consultancy input to ensure that best practices were being incorporated. The team tried to model the behaviours and performance levels being incorporated into the new process and were able to 'test' them safely in this way.

The first key deliverable was the design of the stage gate process from which all other areas of scope would be developed (Figure 10-4). This was based on a typical business change process but included an additional early stage gate so that ideas could be evaluated before much real work had been completed. The aim of this was to encourage generation of ideas at a time when the organization had been used to stability and very little change.

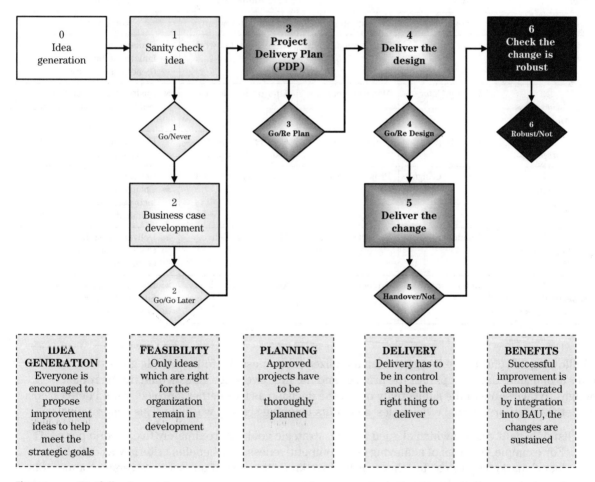

Figure 10-4 Portfolio stage gate process

The stage gate process designed attempted to track key benefits decisions so that benefits for the organization were always being maximized and resources being deployed efficiently. At any stage gate the review process would allow prioritization so that projects could be put on hold should external drivers cause priorities to change or a different balance in the portfolio become necessary.

A part of communicating the role of the management team was the development of a Roadmap Decision Matrix (Melton, 2008) which indicated the decisions required at each stage gate and the data they would expect to be available to support these decisions (Table 10-4).

Table 10-4 Portfolio Decision Matrix

Planning Toolkit – Roadmap Decision Matrix				
Project: Portfolio Development Project			**Sponsor:** Jeff Wilson	
Date: Month 1			**Project Manager:** Trish Bailey	
Stage gate	**Decision**	**Decision by**	**Decision when**	**Data needed**
1 *Idea aligned to business?*	Go or never	Management team	After feasibility work	▶ Benefits of the idea ▶ Link to the business goals (benefits hierarchy)
2 *Business case approved?*	Go or go later	Management team	After project defined	▶ Business case (benefits score and benefits specification)
3 *PDP approved?*	Go or re-plan	Project sponsor	After Project Delivery Plan (PDP) complete	▶ PDP incorporating a benefits realization plan
4 *Design approved for implementation?*	Go or re-design	Project sponsor	After design complete	▶ Design ▶ Change delivery plan incorporating a change sustainability plan
5 *Project scope delivered?*	Handover or not	Project sponsor	After change delivery	▶ Deliverable status report ▶ Customer handover certificate
6 *Sustained delivery of benefits?*	Robust or not	Project sponsor	After benefits delivery	▶ Benefits tracking report ▶ Sustainability tracking report ▶ Customer survey results

Following this part of the design the team realized that unless the four strategic goals could be better articulated, the early part of the process and the first two stage gates would not be robust enough. To take some of the subjectivity out of the decision as to whether a project would deliver 'cost competitiveness' or 'internal efficiencies' a benefits mapping session was held. This generated:

- ▶ A list of benefit criteria which aligned to each strategic goal (approximately five criteria per goal).
  - ▷ For example, in terms of achieving cost competitiveness, one benefits criteria was defined as 'benchmarked services'. In other words, if we know the cost of our services versus competitors we can take actions to become more cost competitive.
  - ▷ This benefits criterion will support achievement of specific service related cost competitive objectives.

- Typical benefits metrics to help Project Teams identify appropriate measures (approximately three benefit metrics per criteria):
  - For example, to measure achievement of service benchmarking there were identified benefits metrics for each facilities management (FM) service: identified benchmark position and improved benchmarked position (in terms of cost position in market).
- An indication of how the various benefit metrics could be scored, based on their contribution to the progress of a specific benefit criterion and therefore a specific strategic goal:
  - For example, service related metrics were weighted according to their contribution to the business profit.
  - The security service is one of 25 separate services within four main FM categories. It brings in 8% of the company revenue and 10% of the company profit even though it represents only 4% of the service portfolio. Any benefits supporting this service gain a weighting appropriate to its profit contribution.
- The basis for a benefits matrix (Table 10-5) which was the foundation for the portfolio spreadsheet that collated data on all the projects in the portfolio:
  - All project ideas were actually evaluated against each of the 20 criteria, although Table 10-5 shows only the summary assessment against each strategic goal.
  - Any idea which could not be easily aligned was dropped.
  - Any idea which indicated sufficient benefits in any one area was developed further until it was developed enough to allow the calculation of a benefits score.
  - The matrix allowed assessment of portfolio balance; preventing all selected projects delivering progress in one strategic area only.

Table 10-5 Extract from 'test' Benefits Matrix

Benefits Management Toolkit – Benefits Matrix

Project:	Portfolio Development Project		Project Manager:	Jeff Wilson
Date:	Month 2		Project Sponsor:	Trish Bailey
Activity	**Benefits criteria**			
	Cost	Efficiency	Quality	Customer
Customer service strategy	None	None	Enabling	Enabling
Lab maintenance improvement idea	To be evaluated	Minor	Improvement in service level	**Major**
Security service strategy	Enabling	Enabling	Enabling	Enabling
Security service internal cost review	**Major**	**Major**	Aim to stay at the same level	Needs to be zero or positive impact
Office services benchmarking review	Enabling	None	None	None
Office cleaning service improvement idea	**Major**	**Major**	Reduction in service level	To be tested
Office management service improvement idea	To be evaluated	**Major**	Minor	Minor

Project Benefits Management

The initial portfolio of ideas tested via the benefits matrix contained over 40 ideas. Of these, 10 were approved for further work, 15 were aligned but low priority and put on hold and 15 were rejected. Table 10-5 contains some of the approved projects. Note that:

- A number of enabling projects were needed as they were the precursors to projects which would deliver significant benefits.
- The projects covered all four strategic areas.
- The total number approved was based on an alignment check and resource review.

Benefits realization

As the project deliverables were completed and the testing phase began, the realization plan was reviewed in more detail and a risk assessment completed so that mitigation plans could be implemented (Table 10-6).

Table 10-6 Portfolio development project Benefits Realization Risk Tool

Benefits Management Toolkit – Benefits Realization Risk Tool

Project:	Portfolio Development Project			Project Manager:		Jeff Wilson	
Date:	Month 4			Project Sponsor:		Trish Bailey	
Benefit	**Probability of not achieving**	**Impact of not achieving**	**Ability to detect failure early**	**Sustainability threat**	**Risk priority**	**Mitigation**	
Projects approved based on strategic alignment	1	5	1	3	15	➤ SOP in place and test robustness ➤ Work with management team	
Efficient use of resources on change projects	2	4	2	3	48	**MEDIUM PRIORITY** ➤ May require additional checks/measures	
Balance of change projects per strategic area	1	5	1	3	15	➤ SOP in place and test robustness ➤ Work with management team	
Benefit score average per project	1	5	1	3	15	➤ SOP in place and test robustness ➤ Work with management team ➤ Encourage idea generation	
Financial benefits delivered	2	4	2	2	32	**MEDIUM PRIORITY** ➤ Need early warning of delivery not progressing to plan	
Projects which identify service quality changes	1	4	1	3	12	➤ Encourage idea generation	
Projects which improve customer satisfaction	3	4	1	3	36	**MEDIUM PRIORITY** ➤ Encourage idea generation ➤ Conduct a customer survey	

(Continued)

Table 10-6 (Continued)

Benefits Management Toolkit – Benefits Realization Risk Tool

Project:	Portfolio Development Project			Project Manager:		Jeff Wilson
Date:	Month 4			Project Sponsor:		Trish Bailey
Benefit	Probability of not achieving	Impact of not achieving	Ability to detect failure early	Sustainability threat	Risk priority	Mitigation
Strategic progress	3	5	3	2	90	**HIGH PRIORITY** ➤ Need early warning of delivery not progressing to plan ➤ Need early indication of how launch portfolio is developing
Scoring system						
Probability	1 = low 5 = high	Impact	1 = low 5 = high	Detection	1 = high 5 = low	Sustainability 1 = low 5 = high

The risk assessment identified the importance of key organizational cultural changes outside the scope of the project, highlighting the importance of the sustainability plan. Therefore an additional mitigation plan was the development of specific sustainability metrics for those areas identified as medium or high priority.

The testing phase continued during months 4 and 5 and during month 6 a handover to BAU commenced. At the end of month 6 during the first sustainability review it was clear that the appointed Portfolio Manager was still heavily dependent on the Project Manager and Project Team however as the portfolio pipeline began to be filled and projects started to move through the stage gates the BAU team gained in confidence. The second sustainability review (Table 10-7) shows this upward trend in sustainability.

As a result of the second sustainability review the Project Manager collated some benchmark data so the Portfolio Manager could better assess the project business cases and the associated delivery and realization plans. For small and well bounded changes impacting only one part of the business, the research data indicated that 'best practice' was implementation within 6 months (Figure 10-5).

Table 10-7 Portfolio development project Sustainability Checklist

Project Management Toolkit – Sustainability Checklist

Project:	Portfolio Development Project	**Date:**	Month 7

Project vision

The business will be capable of selecting and delivering the appropriate mix of business changes for the achievement of strategic success

Sustainability review information

Previous sustainability review	Month 6 number 1	**This sustainability review**	Month 7 number 2
Project representative	Jeff Wilson	**Customer representative**	Trish Bailey

Sustainability checks

Check number	Check	Target (sustained change)	Last review	This review
1	Process is being used appropriately	Robustness of management decisions	Initial review meetings required project input	Project Manager still attends the review meetings to support the stage gate decisions
2	Review meetings are being held as scheduled	Meetings duly incorporated into business planning cycle	Meetings scheduled to year end and one held	Meetings already in next year's schedule. Meetings at management and portfolio level have been successfully held
3	Portfolio is being appropriately populated with ideas and projects	Constant flow of ideas and projects in the process	Flow of ideas is slow, 50% failed stage gate 1	Flow of ideas is still slow but more are showing alignment and passing stage gate 1
4	Portfolio is being appropriately resourced and resource levels are used to guide stage gate decisions	Robust process to assign and manage resources	Not in place	Resource histograms are now in place and used during review meetings. A project was recently put on hold as the required resource was needed for a higher priority project
5	Financial and non-financial benefits are being realized	Robust process in place and in use	In place but only just launched	No forecast benefits to be realized for next month
6	Portfolio balance check – in particular reviewing customer area	Good balance in all four areas with priority in customer area	Very few ideas identified to meet customer goals	Need to identify more about customer expectations and look at benchmarking (cost and service levels)
7	Strategic dashboard is used to check strategic progress	Delivery progress according to plan	Dashboard in use but no progress – too soon	It is still too early to check the use of the strategic dashboard

(Continued)

Table 10-7 (Continued)

Project Management Toolkit – Sustainability Checklist				
Project: Portfolio Development Project			**Date:** Month 7	
Sustainability checks				
Check number	**Check**	**Target (sustained change)**	**Last review**	**This review**
8	How management use the strategic dashboard	Robust use as a part of BAU	Management are struggling with pace of change	Management are starting to use the dashboard to decide on the direction of future change and are considering new operational measures in support
Summary comments and next steps				
There is now an agreed disengagement plan so that the processes and tools become fully integrated into BAU. This means that there will be monthly checks against the remaining areas of developing sustainability. Further data is also needed on the time taken to reasonably deliver some of the identified change projects.				
Is the change completely sustained?		*The text of this cell should read 'Yes/No' Yes should be crossed out to indicate that No has been selected.*	**Date of next sustainability check**	Month 8

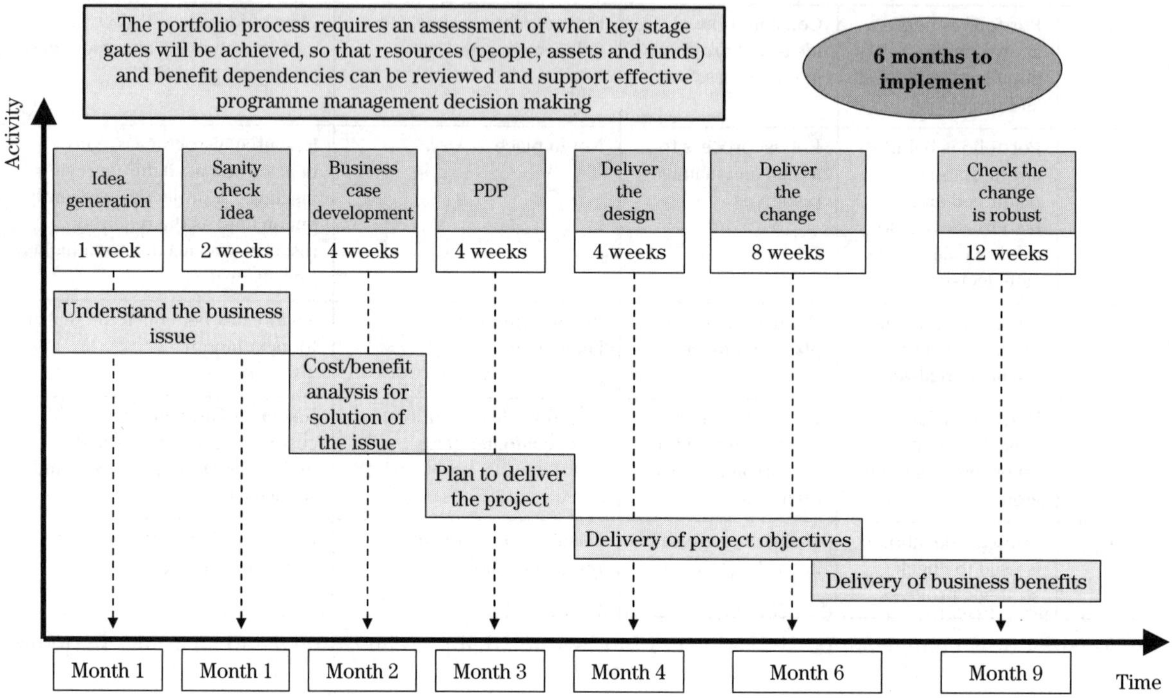

The portfolio process requires an assessment of when key stage gates will be achieved, so that resources (people, assets and funds) and benefit dependencies can be reviewed and support effective programme management decision making

6 months to implement

Activity
Idea generation — 1 week — Month 1
Sanity check idea — 2 weeks — Month 1
Business case development — 4 weeks — Month 2
PDP — 4 weeks — Month 3
Deliver the design — 4 weeks — Month 4
Deliver the change — 8 weeks — Month 6
Check the change is robust — 12 weeks — Month 9

Understand the business issue

Cost/benefit analysis for solution of the issue

Plan to deliver the project

Delivery of project objectives

Delivery of business benefits

Figure 10-5 Portfolio process benchmark data

Following a further 5 months of disengagement work and integration into BAU, the review of the business changes (Table 10-8) supported a formal close-out of the project.

Table 10-8 Portfolio development project 'In Place–In Use' Analysis Tool

Benefits Management Toolkit – In Place–In Use Analysis Tool

Project:	Portfolio Development Project	Project Manager:	Jeff Wilson
Date:	Month 12	Project Sponsor:	Trish Bailey

Benefit deliverable (In Place)	In Place rating	Change required (In Use)	In Use rating	In Place–In Use score
Portfolio process SOP and toolkit (see below)	5	Management team need to use this, facilitated by a Portfolio Manager	4	20
Benefits Matrix Tool	5	Portfolio Manager needs to ensure that this is current and is used for stage gate decisions	4	20
Benefits Scoring Tool	5	Project Managers need to be able to use the tool when developing an idea and a business case	4.5	22.5
Benefits Tracking Tool	4.5	Project Managers need to be able to track benefits on their projects	4.5	20.25
Portfolio scorecard	4.5	Portfolio Manager needs to use this to collate project impact on the business	4.5	20.25
Strategic dashboard	4.25	Management team need to use this to make appropriate strategic decisions	4.5	19.125

Scoring				
In place	1 = incomplete 5 = complete	In use	1 = not in use 5 = fully used	Final score	≥20 = sustained <10 = not sustained

The management team were confident that complete disengagement was appropriate, particularly when reviewing the status of the portfolio and the associated scorecard (Figure 10-6).

During the close-out session all agreed that the benefits of the project had been fully realized and to some extent exceeded. The hidden benefits of changing the culture to one of 'change readiness', innovation in service delivery and a stronger customer focus was a by-product of the new measures system which encouraged all in the organization to:

- Consider how their delivery role:
  - impacted customer receipt of value.
  - impacted organizational economic sustainability.
- Consider how change supported delivery of individual, team, organizational and customer benefits.

Project Benefits Management

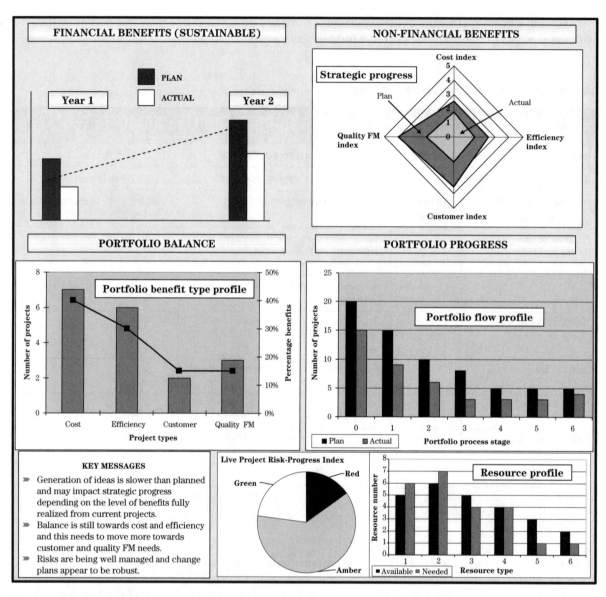

Figure 10-6 Portfolio scorecard

Conclusions

A year after the launch of the new portfolio process the organization was able to report significant growth. This was mainly due to its increase in customer base because of its reputation as an FM provider delivering cost effective fit for purpose services. There is no doubt that the ability to find, prioritize and then deliver the major business changes moved them towards their vision as an FM provider of choice through meeting their strategic goals.

In developing a portfolio process the organization also realized other benefits:

- They developed a programme and portfolio management capability in-house which allowed them to reduce their reliance on external consultants for major programmes or portfolios of change.
- It allowed the organization to develop clearer communication messages at all levels in the organization.
- The management team were able to make better decisions about the use of internal resources (people, assets, funding).

Lessons learnt

- The benefits management processes which we use within project management are appropriate to use within a wider organizational context. Benefits management is an organizational issue.
- The majority of organizations have some portfolio of change in progress at any one time, whether this is through changing business processes, assets or organizational structures and roles. These portfolios need to be managed in a way that aligns the changes (delivered as projects) to the longer term needs of the business, as well as the shorter term operational problems. Benefits management is a way to do this and is a key process within portfolio and programme management.
- Any portfolio of projects needs to have some identifiable link. This can be resources (people, money, things), project objectives or business benefits. Whatever it is, the management of the portfolio should be strongly linked to a benefits management process to aid prioritization.
- Every change in a business should have an identifiable link to the needs of that business.

11 Case Study Four: hurricane preparedness project

This case study has been used because there were very good reasons for both doing, and not doing the project. It provides a demonstration of the need to challenge both the scope and the business environment to establish the true benefits and rationale for the project.

Situation

The business is a data entry facility based in the West Indies, with a predominantly US client base. The company employs approximately 2000 staff. There was a 'near miss' hurricane last summer, which has prompted the CEO to ascertain hurricane preparedness within the company. Several existing clients have also queried the level of preparedness and consequent business risk.

There is a time constraint as it is now January, and hurricane season is generally forecast to start around June. The company is also experiencing significant growth, which is set to continue for at least the next 18 months. The company is located in a business park, very close to a coastal lagoon. There are several other similar local and US 'technology' based businesses here, some of whom have developed preparedness plans, some have not.

If the company is to become prepared, there will be a significant time and cost impact. If a hurricane does not strike, this will appear to have been wasted effort (Figure 11-1). Similarly, if there is a direct

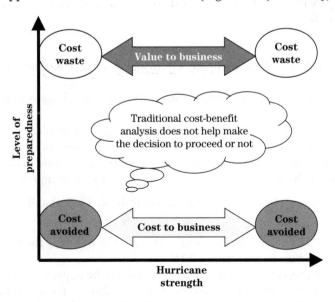

Figure 11-1 Cost-benefits of hurricane preparedness – traditional

169

Project Benefits Management

hit by a category 5 (the strongest category) it is likely that the infrastructure of both the company and the area would be so compromised that the event would be declared a 'force majure' and the business ceased: again a wasted effort. It appears then, that unless a hurricane of a certain strength and size strikes, there is no tangible benefit to being prepared. Indeed, a significant number of businesses in the Caribbean and the US take a view that due to the unpredictable nature of hurricanes, the 'cross our fingers and hope for the best' approach will suffice.

Business case

The core principle of this project is whether benefits of implementing preparedness outweigh the costs – particularly given that there is a high likelihood that the expenditure would either be wasted (no hurricane strike) or of no value (category 5 strike). In order to define the risks more clearly a business case needed to be developed. This would enable the 'idea' of hurricane preparedness to be developed and challenged, providing the first Go/No Go stage gate in the development of the 'hurricane preparedness project'.

The 'Why' Checklist Tool was used to expand on what was, at that point, a gut feeling (Table 11-1). The idea was that the intangible 'soft' benefits (the way being prepared makes us feel) was underestimated, and the apparent 'hard' benefit (if it did ever happen we should be able to save assets) was not enough to justify the expense of the project to the board. After all, hurricanes are not unusual in the region, and the company would already be prepared if the decision makers believed it was a cost effective exercise.

The project was approved on the basis that additional scope definition would be completed and although a traditional cost-benefit analysis could not be done the Project Team were given clear guidance on the level of expenditure for the anticipated benefits.

Benefits definition

The intangible 'soft' benefits were:

- The project should help retain the company's existing client base as preparedness reduces perceived risk.
- The project should improve the company's competitive edge as preparedness reduces risk and increases credibility.
- The project should help recruit and retain staff, as the company would be seen as a good corporate citizen, prepared for and invested in employee welfare in the event of a natural disaster.
- Employees feel more prepared, both at work and at home, and morale is improved.
- Preparedness would help the management team reduce stress and time spent preparing.

The 'hard' (cost based) benefits of preparing for a certain strike probability were:

- We would lose fewer assets.
- We would be able to get back-up and running more quickly, reducing the time (and cost impact) between the event and business as usual (BAU).
- We would reduce the preparation lead time required from the appearance of a hurricane to its impact (the shorter amount of time needed the better, as direction and strength can change rapidly).

Case Study Four: hurricane preparedness project

Table 11-1 Hurricane preparedness 'Why?' Checklist

<table>
<tr><td colspan="4">Hurricane Preparedness Project – 'Why?' Checklist</td></tr>
<tr><td>Project:</td><td>Hurricane preparedness</td><td>Project Manager:</td><td>Derek Thistlewood</td></tr>
<tr><td>Date:</td><td>Week 1</td><td>Page:</td><td>1 of 2</td></tr>
<tr><td colspan="4" align="center">Sponsorship</td></tr>
<tr><td colspan="4">

Who is the sponsor? (The person who is accountable for the delivery of the business benefits)
David Collins

Has the sponsor developed an external communication plan? (How the sponsor will communicate with all stakeholders in the business)
- Weekly face to face preparedness team meetings
- Fortnightly updates to the board
- Kick-off circulation (intranet, department notice boards) to all staff members with project outline
- Completion circulation (intranet, department notice boards) to all staff members. This will include a broad outline of high level company measures, and detailed information of all preparedness issues concerning staff members with meeting/training schedules.

</td></tr>
<tr><td colspan="4" align="center">Business benefits</td></tr>
<tr><td colspan="4">

Has a business case been developed?
This document is serving as the business case in this instance.

Have all benefits been identified? (Why is the project being done?)
- We would loose fewer assets
- We would be able to get back-up and running more quickly, reducing the time (and cost impact) between the event and BAU
- We would reduce the preparation lead time required from the appearance of a hurricane to its impact (the shorter amount of time needed the better, as direction and strength can change rapidly)
- The project would help recruit and retain staff, as the company would be seen as a good corporate citizen, prepared for and invested in employee welfare in the event of a natural disaster
- Preparedness would help the management team reduce stress and time spent preparing. (We know there will be hurricanes, and regardless of whether or not they strike we can cope with the risk better as we are prepared. We don't have to think too much each time as we have a step by step guide, both for our department and for our homes)
- Employees would be better able to continue BAU during the approach of a hurricane (may be several days) as they feel more prepared, both at work and at home. Productivity is less affected as staff are able to maintain focus on the job for longer
- The project would help retain the company's existing client base as preparedness reduces risk. Account Managers have reported several client queries regarding the level of preparedness within the company. We are running the risk of losing existing clients to competitors who can demonstrate preparedness. Clients who have enquired are clients A, C, E and F
- The project will improve the company's competitive edge as preparedness reduces risk and increases credibility. We have lost out on new business recently. One reason could be that our proposals do not address the issue of preparedness. Including such a plan would increase likelihood of winning new business

Who is the customer? (Identify all stakeholders in the business including the customer)
- Authorizing sponsor – CEO
- Champion – Project Manager
- Customers – clients, staff, board, departments, managers, suppliers (everyone)
- Targets – company departments, staff, managers, suppliers

How will benefits be tracked? (Have they been adequately defined?)
At this stage the intangible and tangible benefits have been outlined but not specified in any great detail

</td></tr>
</table>

(Continued)

Table 11-1 (Continued)

Hurricane Preparedness Project – 'Why?' Checklist			
Project:	Hurricane preparedness	**Project Manager:**	Derek Thistlewood
Date:	Week 1	**Page:**	2 of 2
Business change			
Will the project change the way people do business? (Will people need to work differently?) Once the disaster preparedness team (includes representatives from all company levels and departments) is trained, and staff is aware of the location of information, there is little requirement to work differently			
Is the business ready for this project? (Are training needs identified or other organizational changes needed?) ▶ IT/accounts needs to identify alternate area for back-up data to be stored ▶ Additional physical areas of the company need to be sourced, for example, storage area, area for generator and fuel storage tank, water tanks ▶ Disaster preparedness team should be identified and trained (must live close to site and be able to commit time to preparing the company prior to preparing home) ▶ Due to growth forecasts, additional buildings are being considered for occupation. Suitability for preparedness should be built in to these considerations			
Scope definition			
Has the scope been defined? (What level of feasibility work has been done?) The scope definition needs additional work			
Have the benefit enablers been defined? (Will the project enable the benefits to be delivered when the project is complete?) The scope needs to be clearly linked to the benefits and more work is required			
Have all alternatives been investigated? (Which may include *not* needing the project) ▶ Preparedness level A – full redundancy via use of alternate site ▶ Preparedness level B – significant investment in securing infrastructure, assets and manpower ▶ Preparedness level C – some investment in securing infrastructure with increased time investment in securing assets and manpower ▶ Preparedness level D – time investment in securing assets and manpower only ▶ NOT doing the project is the current status quo of the company Option C is recommended. The company has significant budgetary constraints, however, it is vital to protect both assets and manpower.			
Have the project success criteria been defined and prioritized? Based on this review the scope needs to be better defined and then challenged			
Stage One decision			
Should the project be progressed further? (Is the business case robust enough for detailed planning to commence?) Yes – although budgetary constraints will have to be considered with each CSF			

The actual business case based on Figure 11-2 demonstrated that if a hurricane were to occur the value of the plan in meeting the soft benefits, such as employee well being, far out ranked the cost-benefit analysis.

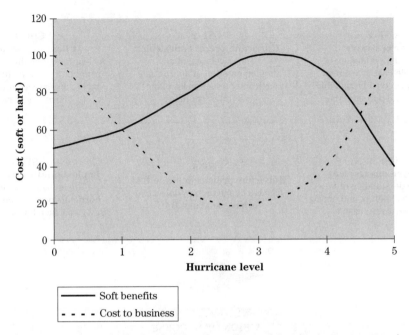

Figure 11-2 Cost-benefits-hurricane preparedness project business case overview

Scope definition

The checklist (Table 11-1) and the improved business case (Figure 11-2) enabled the scope of the idea of preparedness to be more clearly defined (Figure 11-3). The effects of a hurricane can vary from 1 or 2 days of rain to a fully fledged catastrophe, (depending on size, strength, direction and quadrant). In forecasting and planning terms being prepared lies with a scenario somewhere in-between. The level of preparedness for the 'in-between' scenario is dictated to a large extent by budgetary constraints.

This enabled the team to define the following areas of focus. (The '$' indicates those areas that would require capital investment) in order to develop the plan (critical success factor (CSF) 2).

- Preparedness team:
  - Team identification.
  - Preparedness plan outline.
  - Preparedness team communication.
  - Preparedness team training.
- IT:
  - Redundancy ($).
  - Asset protection.
  - Back-ups.
  - Additional hardware/software ($).
  - Department preparedness training.
- Facilities:
  - Redundancy ($).
  - Infrastructure protection ($).
  - Stores.
  - Security.

Project Benefits Management

CSF 1 **Business strategy** A clearly understanding of reasonable cost investment versus tangible and intangible benefits and potential risk	CSF 2 **Hurricane preparedness plan** The development and implementation of appropriate hurricane preparedness plan (IT, facilities, training)	CSF 3 **Client engagement** A clear understanding of what the current and potential future clients expect of a service provider in a hurricane region
CSF 4 **Employee engagement** A clear understanding of how employees feel before and during the hurricane season	CSF 5 **Hurricane preparedness in BAU** The hurricane preparedness plan is integrated into BAU	CSF 6 **Implementation assessment** The impact of having the hurricane preparedness plan is reviewed each hurricane season

VISION OF SUCCESS
Organization is in a state of hurricane preparedness to support sustainabilty of company business (Clients, employees, reputation, assets) within practical limits

Figure 11-3 Hurricane preparedness path of CSFs

- Transportation.
- Canteen.
- Department preparedness training.
- Communication:
  - Client communication.
  - Staff communication.
  - Board communication.
- Accounts:
  - Back-ups.
  - Alternate storage site ($).
  - Department preparedness training.
- Clients:
  - Information collection and comparison.
  - Any special considerations.

The scope of the project was then challenged using the Scope Challenge Checklist (Table 11-2).

This analysis provides two interesting points for further evaluation. The IT infrastructure spend is partially to mitigate against hurricane damage, but also serves a potential purpose as an expansion of capacity into a remote location. The rationale for client benefit is really a feeling that being seen as a good corporate citizen would help the business.

This process resulted in a green light to proceed with a detailed monetary costing analysis, and with those aspects of the plan that involve time and resources, but not capital expenditure. These included,

Table 11-2 Scope Challenge Tool

Benefits Management Toolkit – Scope Challenge Checklist

Project:	Hurricane Preparedness	Project Manager:	Derek Thistlewood
Date:	Week 1	Page:	1 of 1

Scope item	Benefit area	Rationale
Preparedness team	Core team to manage the preparedness process and respond to hurricane events	Provides a single point of reference for actions
IT	Ability to rapidly recover from hurricane strike	Infrastructural spend – expansion of business requires a growth in IT capability – could also provide back-up in the event of hurricane
Facilities	Ability to withstand the effects of a strike and rapidly return the business to operation Employees can stay at work longer and return to work quicker	Facility needs to be able to withstand minor damage and be fully operational within 1 week
Communications	Staff and clients aware of situation Employees not as distracted during hurricane season	Part of the preparedness activities
Accounts	Ability to recover from hurricane strike Employees back in work quicker	Cash-flow and staff could be affected if this area is not back in operation rapidly
Clients	Confidence in preparedness of facilities	Unclear – perceived as of benefit in selling the company to clients as a 'good supplier'

for example, the development of a plan detailing how we will communicate with our clients and our staff, identifying a disaster preparedness team (this is not the Project Team, this is the team that will actually implement the preparedness plan in the event of a hurricane).

The concept of a specific project to become hurricane prepared had not been planned for in the current year's budget. It was therefore decided that based on the strong project case and timeline constraints, the project should go ahead, but with the level of preparedness to a large extent determined by the available funds.

The company was structured so that senior managers reported directly to the CEO. These managers represented each client department, IT, accounts and facilities. All senior managers were expected to take on projects as and when they could. Once a volunteer was selected, he identified a Project Team comprising of representatives from all areas of the business (IT, facilities, client departments, and accounts).

Preparedness team

In order to identify a preparedness team the Stakeholder Management Plan was used. The plan was used in an unusual way as the final outcome of being prepared is to create a 'project in waiting'. Sponsors, stakeholders, champions and targets are part of the project, but will also be critical to the success of any implementation that results from the preparedness project.

From this it became apparent that the whole company and several of its suppliers would be affected by the implementation of this project in some way. An effective preparedness team should therefore comprise representatives of every business area (other than clients and suppliers). These team members would act as change agents who would be involved in becoming prepared for, and trained for implementation. They must be prepared to commit time outside of work both during the project and in the event of a hurricane strike. In real terms this means that any volunteer should live within easy reach of the facility, and be prepared to commit time to preparing the company for a hurricane over and above any outside commitments.

A check list was developed:

- Current telephone contact list to be maintained with alternate numbers in case either cell/mobile or land line service is lost.
- Training in department asset protection.
- Time to be identified outside of working hours to ascertain length of time required to prepare each department.
- Communication hierarchy and systems to be identified.

A final step at this point was to use the Business Environment Checklist (Table 11-3) to evaluate whether the project (which is effectively a change in ways of working for the disaster scenario) would be sustainable in the business environment. The ability to integrate the plan into BAU is a CSF for this project (Figure 11-3) and a part of this involves assessing the business environment. The results indicated a good state of business readiness with some additional actions for the team.

Following this review the project entered the delivery phase with the development of the plan and then integration into the business.

Table 11-3 Business Environment Checklist

Benefits Management Toolkit – Business Environment Checklist			
Project:	Hurricane Preparedness	**Project Manager:**	Derek Thistlewood
Date:	Week 5	**Page:**	1 of 2
Checklist item		**Comment**	
1. Is a champion supported by a dedicated team in place to drive the change?		Yes – key sponsor reporting to the CEO and board	
2. Is the purpose and direction of the change clearly articulated?		Not completely – more definition required of the expected impact on people and ways of working	
3. Are the change drivers understood?		Yes	
4. Is there agreement from key stakeholders on the method and approach?		Partial – some capital spend (not approved for this year) is required which limits what the project can do	
5. Have you communicated the logic behind the change clearly and concisely?		In progress – more definition of the purpose of the change required	
6. Is there mutual trust and understanding?		Yes	
7. Are there communication mechanisms and a means to engage in open dialogue?		Yes – reporting to board level and all key departments engaged	
8. Are people clear about their responsibilities for the change?		Yes	

(Continued)

Table 11-3 (Continued)

Benefits Management Toolkit – Business Environment Checklist

Project:	Hurricane Preparedness	**Project Manager:**	Derek Thistlewood
Date:	Week 5	**Page:**	2 of 2

Checklist item	Comment
9. Is the organization effectively organized to deliver the change?	Yes
10. Do the plans envisage quick wins to help build consensus and maintain progress?	Yes – due to capital constraints, small changes to operational work and facility will help show the direction of change
11. How will those who lose or are damaged by the change be supported?	Main impact will be on those people tasked to secure the facility in the event of hurricane warning – appropriate reward plans need to be developed
12. How will you adapt to changes?	Project review at board level on a monthly basis

Delivery

The departments with the largest potential cost impact were IT and facilities and the areas to address were broken down as follows:

- IT:
  - Redundancy in data systems.
  - Server room equipment redundancy and protection.
  - Production floor asset protection.
  - Support department equipment protection.
- Facilities:
  - Alternate power supply.
  - Alternate water supply.
  - Redundancy in the building.
  - Protection of the building.
  - Protection of workstations, carpets and other infrastructure assets.
  - Development of a storage facility and stocking with supplies (bin bags, toilet roll, food and drink, fans, batteries, radio, walkie-talkies, torches)

Cost-benefit analysis

The IT and facilities representatives within the team consulted with their departments and suppliers and came up with options for each of the areas. Risk assessments were used to judge the impact of each solution on the level of preparedness. The example used for this case study is that of the protection of the building. There are several ways to mitigate the impact of a hurricane on an existing building:

- Install hurricane straps on the roof.
- Install hurricane shutters and/or some form of strengthening of the windows.
- Install lightning rods on the roof.
- Inspect for smaller building leaks and address prior to hurricane season.
- Ensure that there are no smaller objects (e.g. planters) nearby that may be picked up by the wind.

Naturally, those options requiring capital expenditure presented the largest scope for cost variation. In the example of protecting the windows, several cost options were available and are listed in ascending order of cost:

- Apply tape to the window glass so that if it breaks it will not splinter into dangerous shards. This method also strengthens the glass pane. Tape would be kept in stock and applied when a hurricane is deemed a threat.
- Apply hurricane strength tinting to the windows which strengthens the glass pane. This method requires tinting to be applied as part of being prepared.
- Construct plywood shutters for the windows with metal wall brackets. This method requires plywood to be purchased, measured and cut to each window size. Metal brackets are then installed, and the shutters are then put up when a hurricane is deemed a threat.
- Install metal accordion hurricane shutters to each window that are simply drawn shut when a hurricane is deemed a threat. This method requires contracting an outside supplier.

A Benefits Influence Matrix (Table 11-4) shows the way in which each one of these items supports the benefit criteria.

Table 11-4 Benefits Influence Matrix

Benefits Management Toolkit – Benefits Influence Matrix

Project:	Hurricane Preparedness		Project Manager:	Polly James
Date:	Week 1		Project Sponsor:	David Collins

Benefit area	Business process impact			
	Tape	Tinting	Plywood	Metal
Core team's ability to manage the preparedness process and rapidly respond to hurricane events	Low	Medium	Medium	High
IT systems ability to rapidly recover from hurricane strike	Low	Low	High	High
Facility able to withstand the effects of a strike and rapidly return the business to operation	Low	Low	High	High
Staff and clients aware of situation	Low (not visible)	Medium – visible, but not high profile	High	High
Accounts ability to recover from hurricane strike	Low	Low	High	High
Client confidence in preparedness of facility	Low	Low	High	High

This analysis shows that the metal shutters offer no significant business benefit advantage over the plywood ones and are much cheaper to procure. The disadvantage of plywood is the maintenance cost and the time required to 'board up' in the event of a hurricane.

Each requirement was developed in a similar way and the following decisions were made and implemented:

IT

- Where possible redundant capacity data systems were installed (60 day lead time). These ensured functionality if there was damage to the current cables.
- Additional capacity server room equipment was purchased and kept at an alternate site. The protection of existing equipment would be provided by 'bagging' the items and storing them in raised water tight containers.
- Production floor and support department equipment would be protected by 'bagging' the items and raising any that were on the floor (e.g. UPS devices/PC towers).

Facilities

- The company had already installed an alternate power supply due to regular power outages. A feasibility study was carried out to ascertain whether this generator and diesel tank would be sufficient for hurricane preparedness needs. It was found that it was, (assuming no air conditioning was utilized) although a regular testing and fuel top up 'roster' was recommended to avoid a

Project Benefits Management

premium prices/supply issue in the event of a hurricane. The company would be able to run on a full tank for approximately 1 week.
- An alternate water supply was installed. Full tanks would be sufficient to provide drinking and flushing water for approximately 1 week.
- An alternate site was dismissed at this time due to the significant cost.
- Hurricane straps and lightning rods were ordered for the roof, along with plywood shutters for the windows. Remedial works would be carried out on the building to reduce the likelihood of water leaks. The building surrounds would also be kept tidy as part of best practice to reduce the risk of damage by flying objects.
- It was decided that no additional protection of workstations, carpets and other infrastructure assets would be needed, as measures to protect the building should suffice.
- The development of a storage facility was approved with a stock of supplies. These were to include bags for every piece of IT equipment, toilet roll for 50% staff capacity for 1 week, food and drink for the preparedness team (suppliers would be consulted for 50% staff capacity requirement and is discussed later), two fans for each production floor and support area, batteries for radios and walkie-talkies, a radio, hardware items (e.g. screws, drill bits), sandbags, 'wet vac' water vacuum and space to store the shutters.
- Several other facilities issues that involved outside suppliers needed to be addressed:
  - *Transport*
    The company currently utilizes an outside vendor to provide shuttle bus transportation from the site to a central drop off point. In the days after a hurricane, assuming the building is functional, this transport schedule may need to be extended. Currently staffs use local taxis that may not be able to function during lengthy periods of rain. The Project Team proposed an extended schedule covering opening and closing times for all departments.

    Supporting analysis was conducted to ascertain where employees lived and their likelihood of being able to get to the main road. It was decided that enough staff would be able to walk to these roads in order to at least fill 50% of the seats. The transport vendor agreed with this plan on the assumption that fuel would be available on the island. A schedule of testing was developed.
  - *Canteen*
    The company currently utilizes an outside vendor to provide a canteen service, 7 days a week. In the days after a hurricane assuming the building is functional, some form of canteen schedule will need to be preserved. Assuming 50% of the staff are able to get to work, they will probably consume more than normal as they will have a limited supply at home. The primary concern with this vendor is the functionality of their off-site location after a hurricane. A cost sharing scheme for a facility upgrade was agreed. This would provide back-up power and water. A schedule of testing of the off-site facility was developed.
  - *Security*
    The facilities manager secured the stores area and regular checks were conducted as part of BAU. The current security company was also involved at a supplier level so that they had an understanding of requirements in the event of implementation of the plan.

Accounts

The team worked on several areas of preparedness in relation to the accounts department:
- Data back-ups.
- Alternate data storage site.

- A project budget.
- A contingency budget estimate in the event of a hurricane (to include an emergency fund to assist employees with house repairs).

Data back-ups were implemented almost immediately as good practice, with a site on higher ground identified as the alternate data storage site. Rental costs were avoided as space in the CEO's house was available for use at no charge!

The project budget development, and contingency budget estimate was a key area in all aspects of preparedness development, so work was ongoing during this phase of the plan.

Client management

Each of the company's clients had different specialized needs. For example, a large client had vendors at several sites worldwide, so that in the event of call interruption at one site, they could re-route business to other destinations. Another smaller client had only this site, therefore in the event of interruption business would cease.

Despite these differing client needs a generic schedule of communiques was be drawn up along with specific time frame guidelines. This would introduce a pro-active schedule and mitigate against communication confusion in the event of a hurricane. This schedule would form the framework of preparedness in client communication.

A summary was also drawn up with each client's needs, so that support departments (IT, facilities) could prepare for each client's expectations, and a framework for each department's work/no work shut down decision prepared. Data was collected as follows:

- Emergency contact information and hierarchy.
- A specific needs summary for each client.

In addition the team felt that a formal 'Service Level Interruption Plan' (SLIP) should be drawn up in order to set expectations and performance targets in relation to service interruption. This SLIP would be incorporated into existing client contracts and built in to future tenders for business. In doing so, the company would be seen to be behaving in a proactive manner.

Communications

Clients

Client communication has been covered in the client management section.

Staff

Pro-active staff communication was essential, particularly regarding the delivery of 'soft' benefits. These benefits would be:

- Staff confidence in the company (and therefore retention) would improve, as they would feel the company has their welfare in mind and would be prepared in the event of a hurricane.
- Staff morale (and therefore retention) would improve, as sections of the plan (particularly in the set-up of an 'emergency welfare' contingency fund) set the company apart from others in the area.
- Company image within the community would improve, resulting in an increased competitive edge within the local job market.

Staff communication would be provided in a number of ways, all of which would be implemented pro-actively as preparedness measures at the start of the hurricane season:

- At the start of each hurricane season a check list for preparedness in the home would be published on all department notice boards. This would encourage employees to take personal steps to become prepared, increasing their (and their family's) safety and reducing the time they may be away from work in the event of a hurricane.
- A synopsis of the preparedness plan would be published, which would include the set aside of an 'emergency welfare contingency fund' for staff disaster relief. In addition, it would contain a responsibility hierarchy with management contact details for each department.
- Details of how to contact the company in the event of a shut down would be published. For example, which radio stations to tune into for return to work bulletins, land line and cell phone contact numbers.
- The extended transportation plan would be published, giving staff information on where the bus routes would run, and at what times, in the event of its implementation.
- Each senior manager would also be required to keep an up to date list of all staff contact numbers, preferably a land line and cell/mobile phone number in case of loss of either service.

Board

The board would be kept abreast of the preparedness plan implementation via the CEO. He would report on a fortnightly basis.

At the end of the delivery stage the following was the status of each CSF:

CSF 1 – business strategy – defined and agreed.
CSF 2 – hurricane preparedness plan – developed.
CSF 3 – client engagement – achieved.
CSF 4 – employee engagement – achieved.
CSF 5 – hurricane preparedness in BAU – complete but not tested.
CSF 6 – implementation assessment – a part of the benefits realization plan which can only be tested after the first hurricane season.

At this stage the hurricane season was about to commence and the hurricane preparedness plan was about to be tested.

Benefits realization

The Project Manager had done all that was possible in readiness for the next hurricane season and as a part of the project close-out reviewed the benefits for doing this project in the first place (Table 11-5).

Table 11-5 Hurricane preparedness 'Benefits Realized?' Checklist

Project Management Toolkit – 'Benefits Realized?' Checklist			
Project:	Hurricane Preparedness	**Project Manager:**	Polly James
Date:	June 1	**Page:**	1 of 2
Stage Three check			
Have there been any changes since Stage Three Completion? (Note only the changes since the final Stage Three 'health check' No significant changes to the project scope or approach			
Business benefits			
Has the business case changed since Stage One (e.g. during planning and delivery, pre- or post-project approval) The business case remains the same as that approved. Availability of capital limits cost intensive activities			
Have all benefits been defined in terms of trackable metrics? (Why is the project being done?) Major benefits from this project are associated with client and employee image. These have not been formally tracked as metrics for the project, but will be assessed as part of a customer and employee survey post-project execution. Note that no baseline exercise was undertaken for these.			
What is the customer feedback? (Feedback from all stakeholders in the business including the customer) Informal feedback from the CEO and board is that the project has delivered a well understood strategy for dealing with hurricane strikes			
Are the benefits being tracked? No tracking table is defined for the project however employee and client surveys are planned both during and immediately after the next hurricane season.			
Business change			
Is the business ready for the project? (If the project can only enable benefits delivery by changing the way people work – has this been delivered, for example training?) Yes – the business has been full communicated to with regard to the scope of the project, the facility changes and the hurricane preparedness team			
Scope definition			
Has the scope been delivered? Yes			
Have the benefit enablers been delivered? (Are you sure that the project will enable the benefits to be delivered now the project is complete?) Yes – training and support for staff. Communications to key clients on our disaster recovery approach to keeping the business running			

(Continued)

Table 11-5 (Continued)

Project Management Toolkit – 'Benefits Realized?' Checklist			
Project:	Hurricane Preparedness	**Project Manager:**	Polly James
Date:	June 1	**Page:**	2 of 2
Stage Four decision			
Has the project been delivered? (Delivery of project critical success criteria) Yes			
Have the business benefits been delivered? (Why was the project done in the first place?) Cannot be fully tested until after the next hurricane season.			

As a part of the close-out of this project the following operational measures were put in place. These will support measurement of the benefits as developed in the business case (Figure 11-2).

- *Asset recovery cost*: the cost spent on assets due to a hurricane related event.
- *Efficiency*: the availability of employees, seen as an indication of their engagement with the business and their 'well being'.
- *Account downtime*: the level of business disruption caused by a hurricane.

Through review of previous data in the business a baseline level against each measure was possible. In the first season there was a hurricane and the plan was implemented. The efficiency figures did show an interesting trend (Figure 11-4) which, in part, proved that the project had delivered benefits.

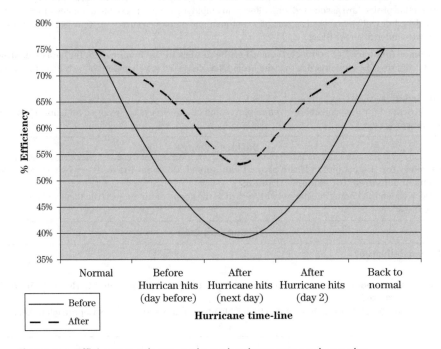

Figure 11-4 Efficiency trends pre- and post-hurricane preparedness plan

Although the expectation of the hurricane still produced a drop in uptime it was significantly less during the recent hurricane than the previous one. This in itself represents revenue into the business that is above normal standards but also reinforces the impact of focusing on softer benefits which initially appear intangible. Another soft benefit which showed an upward trend post-project was employee retention: last season the retention was 94% and this season it had increased to 96%, one of the highest in the region.

The hard benefits were also realized as assets were protected from water damage, facilities were able to provide services to returning employees, connectivity was restored faster than before and accounts were disrupted much less than other local companies.

Of particular interest to the management team was the positive feedback from the clients after the services were resumed post-hurricane. As a result of an increased reputation in the industry additional clients were sourced.

Conclusions

This project illustrates the advantages of using appropriate benefit and scope identification and challenge tools for projects where there is either a very 'soft' benefit or a marginal case for implementation. The likelihood of a hurricane strike on the facility was low, coupled with the fact that a really severe storm would compromise the facility anyway. However, the intangible benefits of putting in place disaster mitigation plans assisted in presenting the company in a better light to clients and staff and aided business growth.

Lessons learnt

- Real benefits do not have to be 'hard' financial numbers. Understand what these benefits are and how they impact the business. Use tools which map these benefits against the overall strategic aims of the business.

- Projects which have substantial infrastructure content (essential for 'image' projects) are particularly susceptible to vanity content. Ensure that the design and implementation scope decisions are regularly challenged against the benefit criteria.

12

Appendices

Appendix 12-1 – 'Why?' Checklist

Project Management Toolkit – 'Why?' Checklist

Project:	<insert project title>	Project Manager:	<insert name>
Date	<insert date>	Page:	1 of 1

Sponsorship

Who is the sponsor? (The person who is accountable for the delivery of the business benefits)
<insert the name of the person who is taking this role>

Has the sponsor developed an external communication plan? (How the sponsor will communicate with all stakeholders in the business)
<insert any comments on how the sponsor was/is communicating with the business>

Business benefits

Has a business case been developed?
<insert comments on the current status of the formally developed business case which supports the project>

Have all benefits been identified? (Why is the project being done?)
<insert comments on the progress of the articulation of the benefits of completing the project>

Who is the customer? (Identify all stakeholders in the business including the customer)
<insert comments on the completion of the stakeholder analysis>

How will benefits be tracked? (Have they been adequately defined?)
<insert comments on benefits metrics>

Business change

Will the project change the way people do business? (Will people need to work differently?)
<consider if the project will change the way that 'normal business' is conducted>

Is the business ready for this project? (Are training needs identified or other organizational changes needed?)
<consider what else is being done in other parts of the business related to the project>

Scope definition

Has the scope been defined? (What level of feasibility work has been done?)
<insert comments on the accuracy of the scope of the project>

Have the benefit enablers been defined? (Will the project enable the benefits to be delivered when the project is complete?)
<insert comments on how the scope is linked to the business benefits>

Have all alternatives been investigated? (Which may include *not* needing the project)
<insert comments on all alternatives to this project which have been considered>

Have the project success criteria been defined and prioritized?
<consider the areas of scope which the project requires to be completed in order to deliver the business benefits>

Stage One decision

Should the project be progressed further? (Is the business case robust enough for detailed planning to commence?)
<insert the decision – yes or no – with comments>

Appendix 12-2 – 'Benefits Realized?' Checklist

Project Management Toolkit – 'Benefits Realized?' Checklist

Project:	<insert project title>	**Project Manager:**	<insert name>
Date:	<insert date>	**Page**	1 of 1

Stage Three check
Have there been any changes since Stage Three completion? (Note only the changes since the final Stage Three 'health check') <Insert comments regarding any changes to the project since the previous health check>

Business benefits
Has the business case changed since Stage One (for example during planning and delivery, pre- or post-project approval) <review the formal business case and confirm its validity – insert any additional comments if there have been changes since approval – note if formal change agreements were made with the sponsor> **Have all benefits been defined in terms of trackable metrics? (Why is the project being done?)** <insert comments with regard to the status of the Benefits Specification Table and whether this has been converted into a Benefits Tracking Table> **What is the customer feedback? (Feedback from all stakeholders in the business including the customer)** <insert comments regarding any feedback from the customer particularly relating to the benefits delivery> **Are the benefits being tracked?** <attach a copy of the Benefits Tracking Table – comment on how long the benefits have been tracked>

Business change
Is the business ready for the project? (If the project can only enable benefits delivery by changing the way people work – has this been delivered, for example training?) <consider how the business has reacted to the changes delivered by the project – attach any completed sustainability reviews>

Scope definition
Has the scope been delivered? <insert comments on the delivery of the project as agreed – refer to the CSFs> **Have the benefit enablers been delivered? (Are you sure that the project will enable the benefits to be delivered now the project is complete?)** <consider if the delivered project enables the benefits to be delivered as agreed by the business case>

Stage Four decision
Has the project been delivered? (Delivery of project critical success criteria) <insert the decision – yes or no – with back-up material such as a completed after action review> **Have the business benefits been delivered? (Why was the project done in the first place?)** <insert the decision – yes or no – with back-up material such as a completed benefits scorecard>

Appendix 12-3 – Economic Evaluation of Projects

Potential projects can be subjected to three types of economic evaluation.

Payback evaluation

Some organizations evaluate projects on a payback basis, calculating the length of time needed to return the initial investment. This has the advantage of being simple but it is not appropriate to most projects.

A simple example illustrates the point. Consider two projects with an investment value of £1 million. The first has a return of £500,000 per year for 2 years and nothing thereafter. The second returns £350,000 every year for 10 years. The former has a payback period of 2 years and the latter something close to 3 years. However, the value of the latter is significantly greater.

Payback can be useful for initial screening and assessment of small short-term projects. For larger or longer-term projects, it is unlikely to provide an adequate assessment.

Discounted cash flow evaluation

The basic methods of assessing the financial merits of projects take account of the time value of money. This represents the effective value of financial returns when a compound interest rate is applied. For example, £10 million now is worth more than £10 million in a year's time, because it could have been invested and earned interest. This is important for projects, because in almost every case, a significant investment is needed during stages One to Three, but revenues are not earned until the project is complete (Stage Four), which could be several years after the investment began.

The time value of money is not the same as inflation. Inflation erodes the value of money and effectively reduces the compound interest rate selected. So, in high inflation periods, a higher interest rate (rate of return) would be required to compensate for the effects of inflation.

There are two, essentially equivalent methods of determining the time value of money within a project. The first, NPV (net present value), takes the scheduled cash flows (at current value) and applies the organization's net cost of capital (expressed as an interest rate) or some other notional rate set as a corporate guideline. These cash flows (adjusted for the time value of money) are then totalled and the result is the NPV. This represents the value of the project to the organization. If the NPV is positive, the organization will be better off doing the project than not. If it is negative, the reverse is true.

In Table 12-1, the project has an NPV of £1,482,934 based on a nominal interest rate of 10% per annum. This means that over 10 years, the effective benefit (difference between money spent and received over 10 years) to the business of investing £5 million now is £1,482,934 in today's money. By way of contrast, the undiscounted cash flow yields a benefit of £5,500,000 – though this is only the equivalent of having £1,482,934 now.

The alternative approach is to calculate the IRR (internal rate of return) of the project. This method determines the notional rate of interest at which the NPV of the project is zero.

It should be noted that it is possible for different projects with different investment levels to return the same value by either method. It is also possible for projects with the same investment levels to return very different results depending on their associated cash flows. It is also possible for a project that pays back its undiscounted capital more quickly to be less valuable to the organization than one with a slower but steadier return. Generally early returns are more valuable than later ones as the discount rate grows geometrically.

Table 12-1 Example NPV

Year	Discount rate	Cash flow	Discounted cash flow
0	1.00	−£5,000,000	−£5,000,000
1	0.91	£900,000	£818,182
2	0.83	£1,000,000	£826,446
3	0.75	£1,100,000	£826,446
4	0.68	£1,100,000	£751,315
5	0.62	£1,250,000	£776,152
6	0.56	£1,250,000	£705,592
7	0.51	£1,250,000	£641,448
8	0.47	£1,000,000	£466,507
9	0.42	£900,000	£381,688
10	0.39	£750,000	£289,157
		NPV	**£1,482,934**

In choosing between projects, an organization should select those with the higher NPV/IRR. For this to accurately reflect the objectives and aspirations of the organization, it will be necessary to include a notional monetary value for any softer objectives. This will need to be done on a consistent and agreed basis to ensure validity and fairness.

Other evaluation methods

A variant of the payback method using discounted cash flows is used by some companies. This is very similar to the IRR/NPV approach and effectively involves solving the cash flow projections for the time at which the project just covers its costs, allowing for the time value of money.

Increasingly, organizations are moving to more sophisticated decision making approaches that use financial modelling to assess the value of a project. Methods similar to those used for valuing financial options may have value in more accurately representing the true value of a project to the organization. There needs to be a balance between the accuracy of the model and the effort needed to analyse the situation. On the other hand, some organizations rely on decision making criteria that take no account of the time value of money such as payback periods or ratios. These are unlikely to be adequate except for the simplest projects.

Where there are significant risk elements in the project, it may also be appropriate to evaluate the range of outcomes against a series of sets of circumstances, either using a sensitivity analysis approach or a simulation method such as Monte Carlo Simulation. When carrying out a sensitivity analysis, it is important to probe the effect of different commercial outcomes as well as cost and delivery issues. Market forecasts on volume and price are also estimates with associated degrees of accuracy. It is critical that the bases of these are probed when scope and cost reductions are being considered.

These methods are inadequate to cover decisions between low risk–low return projects and competing high risk–high return, as there is a need to take the organization's attitude to risk into account. In these cases, it is necessary to consider both the upside and downside risks (potential gains

Project Benefits Management

and the potential losses). The simulation methods can be useful in generating competing scenarios for evaluation.

For projects that are considered 'softer' (for example, where there is no readily quantifiable financial benefit) business case evaluation can often be done on the basis of the relative impact on the business objectives. For example, a change management project can be evaluated by looking at the impact of doing nothing. What will the affect of resistance to change on day-to-day operations be? Using this method, a ranking of 'soft' business cases can be achieved, that will guide the review and approval process.

Taking the same example as was used to generate Table 12-1, Table 12-2 shows the results of a discounted cash flow.

Table 12-2 Example IRR

Year	Discount rate	Cash flow	Discounted cash flow
0	1.00	−$5,000,000	−$5,000,000
1	0.86	$900,000	$772,566
2	0.74	$1,000,000	$736,863
3	0.63	$1,100,000	$695,781
4	0.54	$1,100,000	$597,263
5	0.47	$1,250,000	$582,608
6	0.40	$1,250,000	$500,115
7	0.34	$1,250,000	$429,302
8	0.29	$1,000,000	$294,813
9	0.25	$900,000	$227,762
10	0.22	$750,000	$162,927
		NPV	$0

In Table 12-2, the IRR is 16.5% and the decision making criterion is whether this exceeds the organization's net cost of capital. If it does, the project is worth doing.

Appendix 12-4 – 'How?' Checklist

Project Management Toolkit – 'How?' Checklist

Project:	<insert project title>	**Project Manager:**	<insert name>
Date:	<insert date>	**Page:**	1 of 3

Stage One check

Have there been any changes since Stage One completion? (Development of the business case and project kick-off may be some time apart)
<insert any changes that may impact the delivery of the project or the associated benefits>

Sponsorship

Who is the sponsor? (The person who is accountable for the delivery of the business benefits)
<confirm the name of the person who is taking this role>
Has the sponsor developed a communication plan?
<insert a comment on how the sponsor intends to communicate with project stakeholders during the project>

Benefits management

Has a benefits realization plan been developed?
<insert any data related to the schedule for delivery of the agreed benefit metrics>
How will benefits be tracked? (Have they been adequately defined?)
<insert any additional data which further articulates the specific benefit metrics and which align with work completed during Stage One>

Business change management

How will the business change issues be managed during the implementation of the project? (Are there any specific resources or organizational issues?)
<insert specific plans for the management of the business change associated with the project>
Have all project stakeholders been identified? (Review the stakeholder map from Stage One)
<attach the stakeholder analysis work that has been completed>
What is the strategy for handover of this project to the business? (Link this to the project objectives)
<insert specific plans for project handover>

Scope definition

Has the scope changed since Stage One completion? (Has further conceptual design been completed which may have altered the scope?)
<insert details of the further work which may have been conducted prior to project kick-off>
Have the project objectives been defined and prioritized? (What is the project delivering?)
<attach a copy of the prioritized objectives>
<insert an updated list of project CSFs>

Project type

What type of project is to be delivered? (For example engineering or business change)
<insert the type of project being delivered – note that this is a major category>
What project stages/stage gates will be used? (Key milestones for example funding approval, which might be go/no go points for the project)
<insert the project roadmap for the type of project within the organization>

Appendix 12-4 – 'How?' Checklist (Continued)

Project Management Toolkit – 'How?' Checklist

Project:	<insert project title>	**Project Manager:**	<insert name>
Date:	<insert date>	**Page:**	2 of 3

Funding strategy and finance management

Has a funding strategy been defined? (How will the project be funded and when do funds need to be requested?)
<insert the funding request requirements – estimate accuracy, funding timeline, authorization process>
How will finance be managed?
<confirm that no additional reporting outside of the project control strategy is required>

Risk and issue management

Have the CSFs changed since Stage One completion? (As linked to the prioritized project objectives and the critical path through the project risks)
<insert updated critical path of success if available>
Have all project risks been defined and analysed? (What will stop the achievement of success?)
<comment on any high priority risks>
What mitigation plans are being put into place?
<attach a copy of the high priority mitigation plans>
What contingency plans are being reviewed?
<attach a copy of the high priority contingency plans>
<attach a copy of the Risk Table and Matrix>

Project organization

Who is the Project Manager?
<insert the name of the Project Manager who will be delivering the project in line with the project delivery plan>
Has a project organization for all resources been defined? (Include the Project Team and all key stakeholders)
<insert any comments on the project resource situation – capacity or capability>
<Have roles and responsibilities been defined? Attach the RACI Chart and/or project organization chart>

Contract and supplier management

Has a strategy for use of external suppliers been defined? (The reasons why an external supplier would need to be used for any part of the scope)
<insert a copy of the contract plan>
Is there a process for using an external supplier? (For example selection criteria, contractual arrangements, performance management)
<confirm that procedures to manage supplier selection and performance are in place>

Project controls strategy

Is the control strategy defined?
<comment on each of the following:
- Cost control strategy
- Schedule strategy
- Change control
- Action/progress management
- Reporting
<What methodologies, tools or processes will be used to ensure control?>

Appendix 12-4 – 'How?' Checklist (Continued)

Project Management Toolkit – 'How?' Checklist

Project:	<insert project title>	**Project Manager:**	<insert name>
Date:	<insert date>	**Page:**	3 of 3

Project review strategy

Is the review strategy defined? (How will performance be managed and monitored – both formal and informal reviews and those within and independent to the team?)
<comment on the plan for reviewing project performance during the delivery of the project>

Stage Two decision

Should the project be progressed further? (Is the project delivery strategy robust enough for project delivery to commence?)
<insert the decision – yes or no – with comments on the robustness of the project delivery plan (PDP)>

Appendix 12-5 – Business Case Template

Business Case Template – Title Page

<insert project title>

Project Business Case

Project	<insert project title>
Customer	<insert customer organization>
Date	<insert date of this revision of the business case>
Revision	<insert revision number>
Author	<insert name of Project Manager who has developed the business case>
Distribution	<insert names of those who will receive a copy of the business case>

Business Case Approval

	Project role	Company	Name	Signature	Date
Author	Project Manager	<insert company or specific part of organization>	<insert name>	<author to sign master copy of business plan>	<author to write in date when signed>
Reviewer	<insert role>	<insert company or specific part of organization>	<insert name>	<reviewer to sign master copy of business case>	<reviewer to write in date when signed>
Approver	Sponsor	<insert company or specific part of organization>	<insert name>	<sponsor to sign master copy of business case>	<sponsor to write in date when signed>

Appendix 12-5 – Business Case Template (Continued)

Business Case Template – Table of Contents

<insert project title>

Business Case – Table of Contents

Section number	Section title	Page number
1	**Executive summary**	*<insert page number>*
1.1	The project	*<insert page number>*
1.2	Business background	*<insert page number>*
2	**Introduction**	*<insert page number>*
3	**Project scope and organization**	*<insert page number>*
3.1	Vision	*<insert page number>*
3.2	Implementation strategy	*<insert page number>*
3.3	Deliverables	*<insert page number>*
3.4	Stakeholders	*<insert page number>*
3.5	Related projects	*<insert page number>*
3.6	Organizational impact	*<insert page number>*
3.7	Resources	*<insert page number>*
3.8	Project management framework	*<insert page number>*
3.9	Key issues	*<insert page number>*
3.10	Critical assumptions and constraints	*<insert page number>*
4	**Benefit and cost**	*<insert page number>*
4.1	Benefits	*<insert page number>*
4.2	Costs	*<insert page number>*
5	**Schedule**	*<insert page number>*
6	**Risk**	*<insert page number>*
7	**Alternatives**	*<insert page number>*
7.1	Identification of options	*<insert page number>*
7.2	Comparison of options	*<insert page number>*
7.3	Recommended option	*<insert page number>*
8	**Glossary and appendices**	*<insert page number>*

<Project Title>: Project Business Plan – Revision <insert revision number>

Appendix 12-5 – Business Case Template (Continued)

Business Case Template – Executive Summary

1 Executive summary

The executive summary is the part of the business case that most people will read first, if not the only part. As such, it should summarize the business case, be able to stand alone as a logical, clear concise summary of the business case and highlight the key issues that the reader should be aware of. It should report on the results of the analysis rather than the reasoning behind them.

Things on which to focus:

- The definition of the business needs.
- Relationship to the strategic/corporate plan.
- Summary of options.
- Costs, benefits and key recommendations.

There is no need to include:

- Assumptions and constraints (unless they are key).
- Analysis, reasoning or details of any form (remember that the remainder of the business case will contain all required details).

This should be developed after the rest of the document has been completed.

1.1 The project

Write a brief paragraph on the overall goal of the project – the vision of success. Highlight any issues which came up during the business case development process.

1.2 Business background

Write a brief paragraph on the reason why the project is needed and the business benefit it will deliver. Highlight any critical business change elements which are NOT included in the project but which are critical to success – realization of the business benefits.

Conclude with a brief paragraph explaining the aim of the business case so that readers understand how it will be useful for them. Bear in mind that the business case has a wide audience and different stakeholders will use it in a variety of different ways.

Appendix 12-5 – Business Case Template (Continued)

Business Case Template – Introduction

2 Introduction

This is used to introduce the business need, briefly describe what has happened in the past to address the problem, and then describe the current status at the time of writing the project business case.
Explain why the project business case is being produced. Generally it is to:

- Define the business need or problem in detail.
- Analyse options (where resources have already been allocated this may involve determining what can be delivered with those resources).
- Identify costs, benefits and risks.
- Put forward a proposal to senior management for approval to proceed with the project, or to the funding source for approval for funding of the project.

Ensure that it is clear:

- Who is sponsoring the development of the project business case?
- Who is the document intended for?

<Project Title>: Project Business Case – Version <insert revision number>   *Page <insert page number>*

Appendix 12-5 – Business Case Template (Continued)

Business Case Template – Project Scope and Organization

3 Project scope and organization

The information in this section is important, as it will form the basis of the project business plan if the project/initiative proceeds. It defines the scope of the project.

3.1 Vision

What is the goal of the project, what is it expected to deliver? A high level description of the objective(s) of the recommended option contained in this project business case (a 1-liner).

3.2 Implementation strategy

Identify the type of approach to implementing the preferred option, for example one large project, a number of smaller projects or a combination of both. The breakdown of the projects within this strategy can also be included where the 'manageable chunks' or phases for each project have been identified.

3.3 Deliverables

Complete list of deliverables and their critical fitness-for-purpose features.

3.4 Stakeholders

List of those to whom outputs will be delivered and a short description of how each stakeholder will utilize them to generate target outcomes.

3.5 Related projects

Related projects (or major change initiatives) can be of significance to the project. List related projects dependent on/or interdependent with this project; or whether this project is dependent on other projects. The nature of a dependency can include a shared relationship with data, functionality, staff, technology and/or funding.

3.6 Organizational impact

How will the work undertaken during the project impact on the organization, and how will these impacts be addressed?

3.7 Resources

Budget and list of critical human/information resources – noting when those resources are required for the project. Include such things as: budget, human resources, information, infrastructure required.

<Project Title>: Project Business Case – Version <insert revision number>   *Page <insert page number>*

Appendix 12-5 – Business Case Template (Continued)

Business Case Template – Project Scope and Organization

3.8 Project management framework

This describes how the project will be managed. It is an outline only; the detail should be contained in the project business plan. Outline the proposed project governance model (who is responsible for what) and the plan for organizational change management.

3.9 Key issues

Identify any key issues, why the issues are important and how they will be addressed.

3.10 Critical assumptions and constraints

It is essential that assumptions made during the business plan development process are recognized and recorded. This may include assumptions arising from previous documents, such as a project proposal/brief, a strategic information systems plan or other existing business documents. Include a discussion of the implications if the assumption is wrong. This is often a risk to the project. Any requirements for specialist resources or skills should be identified here, and any dependencies that exist with other projects or initiatives. Constraints may change, so a discussion of the implications of this should also be included. For example, the budget that has already been allocated may not be the constraint it initially appears to be.

<Project Title>: Project Business Case – Version <insert revision number>

Appendix 12-5 – Business Case Template (Continued)

Business Case Template – Benefit and Cost

4 Benefit and cost

This section analyses the recommended option in detail. The benefits, plus the direct and recurring costs should be identified.

Depending on the scope of the organization for which the project business case is being prepared, it may not be appropriate for some financially quantifiable benefits or costs to appear as direct in the analysis, but rather as indirect financially quantifiable benefits.

To justify a recommendation the analysis should incorporate economic and business outcomes to be delivered/received from the proposed initiative. For example, meeting a customer service obligation is an important non-financial benefit for a service delivery organization.

4.1 Benefits

The benefits and issues need to be listed. These should include such things as:

- Hard (cost) related benefits.
  - Increased revenue.
  - Cost reductions, for example, reduced maintenance, reduced staff costs.
  - Cost avoidance, for example, increased service with the same staff.
- Soft related benefits.
  - Service enhancement, for example, faster service, greater availability.

4.2 Costs

Typical costs include capital and recurring costs including human resource costs. It is important to include recurring costs as these will occur after the project has closed and must be budgeted for separately by the organization.

Most costs can be reduced to a currency amount and analysed over time. The appropriate time to review will be determined by the expected life of the change initiative and advice should be sought from the business owner on what is considered appropriate.

The following costs should also be included:

- Risk management costs.
- Quality management costs.

Sources of funds, if required, may also be identified here.

<Project Title>: Project Business Case – Version <insert revision number>   *Page <insert page number>*

Appendix 12-5 – Business Case Template (Continued)

Business Case Template – Schedule, Risk, Alternatives and Appendices

5 Schedule

Outline of project phases, major areas of work and key milestones.

6 Risk

Identify critical risks that will prevent the project delivering.

7 Alternatives

This is a high level analysis of the possible alternatives that could be employed to bridge the gap between the current situation and what is proposed.

If there are a number of significantly different options for proceeding, a feasibility study may have been carried out on one or more of these options that reduces the complexity of identification.

7.1 Identification of options

Generally if a detailed analysis of options is required, fewer significant options are preferable to many. To ensure that a thorough assessment is made of all possible alternatives, a minimum of three options may need to be considered, such as:

- Option 1 – Do nothing.
- Option 2 – An option that would achieve the same result.
- Option 3 – The preferred option.

Any minor variations between options should be resolved at this stage based on the assumptions, constraints, areas of risk (if identified); leading to the best case scenarios for each major option. If the analysis results in only one option appearing viable, then that analysis should still be demonstrated.

7.2 Comparison of options

The benefits, direct and recurring costs and major risks should be identified for each option. This should be a summary and may best be displayed in a table. If a detailed analysis is necessary it should be included in the Appendices.

7.3 Recommended option

The recommended option from the previous analysis should be identified here.

Appendix 12-5 – Business Case Template (Continued)

Business Case Template – Schedule, Risk, Alternatives and Appendices

8 Glossary and appendices

Appendices can help the document flow better, especially during the analysis and justification sections (during the 'argument' parts) by extracting information out of the body of the document for reference. For example, the following may be useful:

- A detailed cost/benefit analysis for each option and any relevant estimates (only if required, for large projects only).
- A risk analysis of the options.
- A glossary, if there are a lot of terms or concepts that are likely to be unknown or confusing (such as acronyms).

<Project Title>: Project Business Case – Version <insert revision number>   *Page <insert page number>*

Appendix 12-6 – Simple Benefits Hierarchy Tool

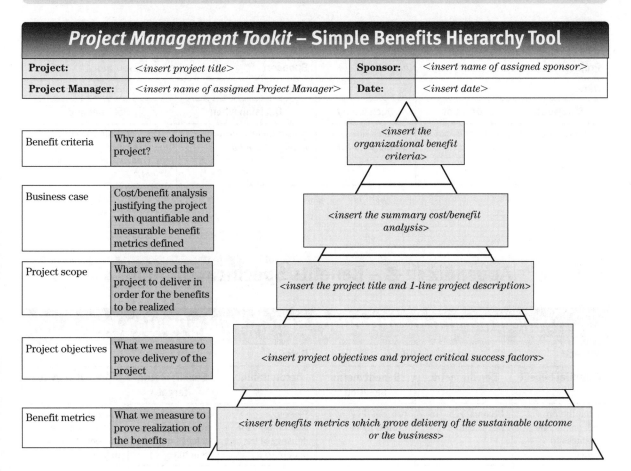

Project Benefits Management

Appendix 12-7 – Roadmap Decision Matrix

Planning Toolkit – Roadmap Decision Matrix

Project:	<insert project title>			Sponsor:	<insert name>
Date:	<insert date>			Project Manager:	<insert name>
Stage gate	**Decision**	**Decision by**	**Decision when**		**Data needed**
<insert number>	<insert decision to be made>	<insert name of decision maker>	<insert when the decision needs to be made>		<insert the data needed to make the decision in terms of meeting acceptance criteria or not>

Appendix 12-8 – Benefits Specification Table

Project Management Toolkit – Benefits Specification Table

Project:	<insert project title>		Date:		<insert date>
Potential benefit	**Benefit metric**	**Benefit metric baseline**	**Accountability**	**Benefit metric target**	**Area of activity**
What the project will enable the business to deliver	Characteristic to be measured	Current level of performance	Person accountable for delivery of the benefit to target	Required performance to achieve overall benefits	The project scope that will enable this benefit to be delivered
<insert the benefit criteria which this benefit metric is related to>	<insert the specific benefit metric which is to be measured>	<insert the measured level of current performance, that is 'before' project commencement>	<insert the name of the person who is accountable for the delivery of this specific benefit>	<insert the target level of the metric at the target date>	<insert the area of the project which is specifically related to the delivery of this benefit>

Appendix 12-9 – Business Case Tool

Project Management Toolkit – Business Case Tool

Project:	<insert project title>	Date:	<insert date>
Business case developed by:	<insert the name of the Project Manager developing the business case>	**Date:**	<insert the date issued for approval>
Project reference number	<insert the project reference number as it will be referred to within the organization>	**Business area**	<insert the business area or organizational group who initiated this project>
Project Manager	<insert the name of the person who becomes accountable for the delivery of the CSFs if the project is approved>	**Project sponsor**	<insert the name of the senior organizational stakeholder who becomes accountable for the delivery of the benefits from this project if it is approved>
Business background	<insert a description of how the potential project was identified, within which area of the business and a summary of the problem statement>		
Project description	<insert brief description of the project – the scope and the CSFs. If necessary attach the Simple Benefits Hierarchy which will have been updated during Detailed Benefits Hierarchy development (do not attach this as it is an internal project tool)>		
Delivery analysis	<insert a summary of the resources required to deliver the project (internal and external), capital and revenue; insert a summary of a high level risk assessment and include any dependency issues related to other 'live' projects or activities within the organization, attach a preliminary milestone schedule>		
Business change analysis	<insert comments regarding the potential impact of the project should it be approved – the impact of this project on the business>		
Value add analysis	<insert the summary of the Detailed Benefits Hierarchy in terms of the high level benefit criteria, the cost/benefit analysis and the specific benefit metrics; discuss any link to key milestones>		
Impact of NOT doing the project	<discuss the level of urgency regarding this project – what happens if the project is delayed or is a lower priority than another? Increased costs? Customer dissatisfaction? Lost customer orders? What is the urgency of the problem which this project will solve?>		
Project approved (Value add or Not?)	Yes/No <approval to proceed to the next project stage gate>	**Name of approver and date**	<insert the name and position of the person approving the project, signed and dated by them>

Appendix 12-10 – Benefits Tracking Tool

Project Management Toolkit – Benefits Tracking Tool

Project:		<insert project title>		**Date:**		<insert date>	
Benefit metric		**Baseline**	**Milestone 1**	**Milestone 2**	**Milestone x**	**Target**
		<insert baseline date>	<insert date or activity>			<insert target date or activity>
Metric 1 <insert metric from benefits specification table>	Plan	<insert baseline data>	<insert planned metric level>			<insert target metric level>
	Actual	<insert baseline data>	<insert actual metric level>			<insert actual metric level>
Metric 2	Plan					
	Actual					
Metric x	Plan					
	Actual					

Appendix 12-11 – Sustainability Checklist

Project Management Toolkit – Sustainability Checklist

Project:	<insert project title>	Date:	<insert date>
Project vision			
<insert brief description of the vision of sustainability>			
Sustainability review information			
Previous sustainability review	<insert date and review number>	**This sustainability review**	<insert date and review number>
Project representative	<insert name>	**Customer representative**	<insert name>
Sustainability checks			

Check number	Check	Target (sustained change)	Last review	This review
<insert check number>	<insert a check which will support the continued sustainability of the change which has been implemented> <note that sustainability checks do not track benefits>	<input a measurable target with appropriate units>	<insert measure, or not applicable if this is the first review>	<note level of measure and insert any appropriate comments>
<insert check x>	<insert as few or as many checks as necessary to assure all stakeholders that the change has been sustained>			

Summary comments and next steps			
<insert comments regarding the results of the sustainability check and any actions agreed with the customer>			
Is the change completely sustained?	Yes/No <delete as appropriate>	**Date of next sustainability check**	<insert date>

Appendix 12-12 – Benefits Realization Plan Tool

Planning Toolkit – Benefits Realization Plan Tool

Project:	<insert project title>	Project Manager:	<insert name>
Date:	<insert date>	Page:	1 of 1

Benefits concept

<insert a copy of the Simple Benefits Hierarchy>

Benefits specification

<insert a copy of the Benefits Specification Table>

Benefits realization

Benefit metric	Tracking frequency	Dependency		Priority versus vision of success
		Project scope	Business change	
<insert benefit metric to be tracked>	<insert when tracking starts and at what frequency>	<insert any scope items, without which the benefit metric will not improve>	<insert any change external to the project, without which the benefit metric will not improve>	<insert criticality of this benefit metric to the overall achievement of the vision of success>

Appendix 12-13 – Glossary

Benefits criteria
The reason the project is being done; the articulation of benefits as linked to an organizational objective.

Benefits enabler
What the project has to deliver to enable the benefits to be delivered; the project scope.

Benefits hierarchy (from *Project Management Toolkit*)
A tool which confirms the alignment of the intended project scope to the targeted business benefits.

Benefit metrics
Those measures which will confirm, after project delivery, that the business is realizing the benefits.

Benefits mapping
A method which identifies and articulates the benefits that relate to the specific organizational goal.

Critical success factors (CSFs)
A quantifiable/measurable and identifiable action/activity that has the potential to impact the overall success of the project.

Critical path of success (from *Project Management Toolkit*)
The linkage of high level critical success factors to form a critical path of quantifiable/measurable actions/activities that have the potential to impact the overall success of the project.

Critical success hierarchy (from *Project Management Toolkit*)
The hierarchy of CSFs which define the scope of the project. With each level in the hierarchy the scope detail increases.

Monte Carlo simulation
Simulation method utilizing repeated generation of random values for unknown criteria.

Near shore
A term describing an outsourcing vendor, that despite being located in a different geographic area (usually country), has a similar time zone/culture to its clients.

Appendix 12-14 – References

Earl, M.J. and Hopwood, A.G. (1980) *From management information to information management.* In: Lucas, H.C. et al. (ed.) *The Information Systems Environment*, North Holland Publishing, The Netherlands.

Eccles, T. (1994) *Succeeding with Change: Implementing Action-Driven Strategies*, McGraw-Hill New York, USA.

Fahey, L., Narayanan, V.K. (1986) *Macroenvironmental Analysis for Strategic Management.* St Paul: West Publishing Co., New York, USA.

Jick, T.D. (1991) *Implementing Change.* Case N9-491-114, Harvard Business School Case, MA, USA.

Kaplan, R.S. and Norton, D.P. (1996) *The Balanced Scorecard: Translating Strategy into Action.* HBS Press, Boston, MA, USA.

Kotter, J.P., Schlesinger, L.A. (1979) *Choosing strategies for change.* Harvard Business Review, 57(2): 106–114, MA, USA.

Lewin, K. (1951) *Field Theory in Social Science.* Harper and Row, New York, USA.

Porter, M.E. (1980) *Competitive Strategy: Techniques for Analyzing Industries and Competitors.* Free Press, New York, USA.

Porter, M.E. (1985) *Competitive Advantage: Creating and Sustaining Superior Performance.* Free Press, New York, USA.

Melton, T. (2007) *Project Management Toolkit: The Basics for Project Success*, Elsevier, 2nd Edition, Oxford, Great Britain.

Melton, T. (2008) *Real Project Planning: Delivering a Project Delivery Strategy*, Elsevier, Oxford, Great Britain.

Kipling, R. (1902) *Just So Stories*, Macmillan, London, Great Britain.

Bicheno, J. (2004) *The New Lean Toolbox: Towards Fast, Flexible Flow*, PICSIE Books, Buckinghamshire, Great Britain.

Index

A
After Action Review (AAR)   108
activity mapping   18, 35
asset recovery cost   184

B
benchmark data   164
Benefit Realization Risk Tool   91–93, 161–162
benefit(s)
  area   59, 79
  concept   16, 29, 31–35
  criteria   12, 29, 41, 46, 55, 76, 158
  delivery   2, 9, 10–11, 131
  enablers   12
  estimation   133, 134
  explicit   18, 93–94
  financial   46–47, 166
  hard   77–78, 170
  implicit   18, 98–99
  management   3, 4, 5, 16, 19
  management lifecycle   16, 17
  map   38, 148
  mapping   31, 33, 35, 38
  matrix   36
  measurement   43, 75, 80
  metrics   43
  multiplier   47
  non-financial   46, 166
  Pareto score   48, 49
  realization   3, 12, 16, 50, 87, 147–148, 161, 183
  realization process   87, 88
  reporting   94–97
  risk assessment   18
  scorecard   44, 46–47
  scoring   43, 48
  soft   78, 170
  specification   3, 16, 53, 69, 75, 76, 80
  sustainability   103
  tracking   9, 16, 18, 91, 94, 97, 147
Benefits Hierarchy Tool   11–12
Benefits Influence Matrix   18, 78–80, 179
Benefits Mapping Tool   18, 36–37, 39–40
Benefits Matrix Tool   18, 41–43, 159
Benefits Realization Plan Tool   18, 210
Benefits Realization Risk Tool   18, 91–93, 161–162
'Benefits Realized?' Checklist   18, 183–184, 189
Benefits Scoring Tool   18, 45–48, 155
Benefits Specification Table   18, 135, 156, 206
Benefits Tracking Tool   18, 147, 208
bottom-up mapping   31
business
  benefits   53
  case   8, 25, 170
  case development   1, 6, 8, 69–85
  change management   13, 63–66
  change plan   11
  environment statement   20
  links   27–28
  need   76
  objectives   53
  process   135
  sponsor   37
  strategy   11–12
  sustainability   106
business as usual (BAU)   1, 93, 103, 162
business case, 12, 15, 76, 87, 120, 151, 170
  content   82–84
  development   1, 6, 7–8, 69, 131
  development cycle   69
  fundamentals   81–82
  one page document   80, 81
  standard   80, 81
Business Case Template   18, 84, 138–145, 196–204
Business Case Tool   18, 81, 112, 119–120, 207
Business Environment Checklist   18, 66, 176—177
Business Satisfaction Analysis Tool   18

C
capital expenditure   74, 178
client management   181
communications   100, 174, 181–182
conceptual design   131, 132–133
consultancy lifecycle   89
contract   89, 99, 100, 107
cost-benefit analysis   18, 57–58, 178
cost estimating   72–75
Critical Success Factor (CSF)   11–16, 43, 53, 152, 174
Critical To Quality (CTQ) criteria   18
CTQ Scope Definition Tool   55–56, 114, 154
customer   99, 171, 188
  contracts   18, 90, 100
  satisfaction   100, 101–102
Customer Satisfaction Analysis Tool   18, 100–101

Index

D
deliverable   8, 11, 56, 133, 139, 200
delivery   1, 2, 9, 133, 146, 157, 178
detailed design   131, 133–136
discounted cash flow evaluation   26, 190–191
disengagement   18, 89, 108

E
earned value   94, 97–98
economic evaluation
   discounted cash flow evaluation   190–191
   methods   191
   playback evaluation   190
end user   49
engagement   89
explicit benefits   93–94, 98–99

F
Failure Modes and Effects Analysis (FMEA)   91
financial benefits score   46
financial cost avoidance   47
Fishbone Diagram   33
'five whys'   5, 8
forward investment   59

G
gaining entry   89

H
hard benefits   77–78, 147, 185
hierarchy of objectives   18, 53, 54
'How?' Checklist   18, 193–195

I
implicit benefits   98–99
In Place–In Use Analysis Tool   104–106, 165
Input–Process–Output (IPO) Diagram   2
internal rate of return (IRR)   127, 190, 192
issues mapping   33

K
Kano analysis   18, 55, 100, 101, 153

L
lean thinking   18, 70
lean value management   18, 71–72
level of influence   79
Lewin's force field model   126

M
milestone report   94–95
mitigation plan   92, 194
Monte Carlo Simulation   191

N
needs–wants matrix   58–59
net present value (NPV)   190
non-financial benefits score   46

O
one-off financial savings   47
operating expense   74
opportunity cost   26–27
organizational strategy   19
   business environment changes   20
   competitive environment   21
   mission   19
   project identification   22–26
   stakeholder expectations   21

P
path of CSFs   152
payback evaluation   190
PEST model   20
Porters Five Forces   21
Portfolio Decision Matrix   158
portfolio development   152, 153, 155–157, 161, 164
portfolio of change   14
portfolio scorecard   166
potential project   61, 115, 116
profile report   95–96
progress against plan   94
project and business
   business links   27–28
   economic evaluation of   26
   opportunity cost   26–27
   organizational strategy   19, 22
Project Delivery Plan (PDP)   62, 83, 87
Project Manager   7, 30, 45, 155, 183, 194, 207
project(s)   13, 24, 79, 106, 118, 126
   approval   131–131
   and BAU   10
   and business   19
   benefit estimation   134
   categories   4
   champion   49
   chaos   6, 9
   closure   18, 108
   completion   107
   cost estimating   73, 143
   economic evaluation of   26, 190
   flow issues   4
   handover   107
   impact of   74–75
   lifecycle   1–3, 74
   matrix   71
   objectives   12
   potential   61, 115, 116, 126
   requirements   98
   review   108

scope 53–59, 200
selection 132
strategic level 108
strategy 62–63
success 23, 109
teams 53, 82

R
RAG report 96
realization planning 90
resistance to change 18, 192
risk management matrix 27
risk priority number 92
Roadmap Decision Matrix 18, 158, 206
root cause analysis 7, 10, 34

S
'satisficing' 63
Scope 17
  benefit analysis 56
  and business change management 63–66
  definition 18, 173
  item 59
  project 53, 55, 56, 82
  and project strategy 62–63
scope-benefit analysis 56
Scope Challenge Checklist 18, 59, 175
Scope Definition Tool 18
sensitivity analysis 191
Simple Benefits Hierarchy 13, 18, 118, 205
soft benefits 78
Specific, Measurable, Appropriate, Realistic and Time-based (SMART) metrics 133

sponsor 37, 48, 85, 171, 193
stage gate process 14–15, 157
stakeholder
  expectations 21
  management 18, 48, 49
STEEPLE model 20
STEP model *see* PEST model
success evaluation 108
Sustainability 9, 18, 50
  checking 103–104
  planning 90
  threat 92
Sustainability Checklist 18, 163–164, 209
sustainable financial benefits 47
SWOT analysis 22

T
top-down mapping 31
triage method 57

U
uncertainty 25

V
value-add analysis 70, 81, 112, 120
value engineering 71
value management 70, 71, 72
value stream 70

W
'Why' Checklist 18, 171, 188